T-Labs Series in Telecommunication Services

Series editors

Sebastian Möller, Berlin, Germany
Axel Küpper, Berlin, Germany
Alexander Raake, Berlin, Germany

More information about this series at http://www.springer.com/series/10013

Jan-Niklas Antons

Neural Correlates of Quality Perception for Complex Speech Signals

Springer

Jan-Niklas Antons
Quality and Usability Lab
Technische Universität Berlin
Berlin
Germany

Zugl.: Berlin, Technische Universität, Diss., 2014

ISSN 2192-2810 ISSN 2192-2829 (electronic)
T-Labs Series in Telecommunication Services
ISBN 978-3-319-38654-6 ISBN 978-3-319-15521-0 (eBook)
DOI 10.1007/978-3-319-15521-0

Springer Cham Heidelberg New York Dordrecht London

Softcover reprint of the hardcover 1st edition 2015

Printed on acid-free paper

Springer International Publishing AG Switzerland is part of Springer Science+Business Media (www.springer.com)

Preface

This book presents the research of the author on the neural correlates of quality perception for complex speech signals. Two different disciplines will be interconnected here, namely neuroscience and Quality of Experience research, which do not seem to be frequently used in combination for research on speech quality perception. In the five experiments conducted here, standard clinical methods in neurophysiology on the one hand, and on the other hand, methods used in fields of research concerned with speech quality perception, will be applied. Using this combination, it will be shown that speech stimuli with different lengths (phonemes, words, sentences and audiobooks) and different quality impairments (signal-correlated noise, reduced bit rate of a speech codec and reverberation) are accompanied by physiological reactions related to quality variations, e.g. a positive peak in an event-related potential. Furthermore, it will be shown that—in most cases—quality impairment intensity has an impact on the strength of the intensity of physiological reactions (components of event-related potentials in Chaps. 2–4, or alpha frequency band power in Chaps. 5, and 6). This book consists of the following contributions: Implementation of a test set-up combining neurophysiological and subjective quality assessment methods for speech quality perception testing (Chaps. 2–6). The proof that this test set-up successfully functions with short speech stimuli (phonemes) and generic quality impairment, i.e. signal-correlated noise (Chap. 2). A successful application of this test method to longer speech stimuli (words) with a more realistic quality impairment, i.e. reduced bit rate of a speech codec (Chap. 3). The proof that this technique successfully functions in respect to stimuli with lengths for standard quality testing (sentences) and an environment-related quality impairment, i.e. reverberation (Chap. 4). An investigation of the impact of a speech compression algorithm with reduced bit rate on the cognitive state of listeners for speech stimuli of long duration (audiobooks) in constant (Chap. 5) and varying quality conditions (Chap. 6).

Berlin, December 2014 Jan-Niklas Antons

Acknowledgments

This work would not have been possible without the help of numerous supporters. Thank you to everyone who supported me during this work. A special mention belongs to the following institutions and persons:

- I am grateful to the Technische Universität Berlin, the Telekom Innovation Laboratories, and the Bernstein Focus: Neurotechologie—Berlin which provided the foundation for all my work. I am especially thankful for the effortful work of Dr. Heinrich Arnold, Prof. Dr. Klaus-Robert Müller, Prof. Dr. Benjamin Blankertz and their teams: thank you for making this work possible.
- I am grateful to the Quality and Usability Lab, the group Assessment of IP-based Applications, and all the colleagues who supported me over the years. You let me experience how it feels to work in a great team.
- I am grateful to my student workers Ahmad Abbas and Steffen Zander who supported me brilliantly over the years.
- I am grateful to Sebastian Arndt, Dr.-Ing. Benjamin Belmudez, Dr.-Ing. Marcel Wältermann, Prof. Dr.-Ing. Alexander Raake, Dr.-Ing. Tim Polzehl, Dr. Benjamin Weiss, Dr.-Ing. Marie-Neige Garcia, and Dr.-Ing. Jens Ahrens with whom I spent much time discussing and working in one on the most interesting research areas.
- I am grateful to Irene Hube-Achter, Yasmin Hillebrenner, and Tobias Hirsch whom organized the Quality and Usability Lab so brilliant over all the years and helped me in every situation with a well-suited solution.
- I am grateful to Dr. Robert Schleicher who supported me from the first day of my postgraduate time in uncountable manners. Thanks for the good time and your true understanding in so many situations.
- I am grateful to the reviewers of my doctoral thesis: Prof. Dr. med. Gabriel Curio and Prof. Tiago H. Falk, Ph.D., for their scientific and thesis-related support over the last years.
- I am grateful to my supervisor Sebastian Möller who supported me not only in every research-related question but also showed me how to organize (business-) life.

- I am grateful to Viktoria Voigt for supporting me in all possible ways: scientific, business and private matters. Thank you for your love and for understanding me. Thanks for pointing me in the right direction when I am too blind to find it on my own.
- I am grateful to my parents for the years of support during my entire life. You are the foundation of my entire life, thanks for being there for me and understanding me. Thank you for teaching me strength and strong will.

Contents

Acronyms

2AFC	Two-alternative forced choice
ACR	Absolute category rating
ANOVA	Analysis of variance
AUC	Area under the curve
BCI	Brain-computer interface
CCR	Comparison category rating
CQS	Continuous quality scale
DCR	Degradation category rating
DSCQS	Double-stimulus continuous quality scale
EEG	Electroencephalography
ERP	Event-related potential
EU	European Union
FBP	Frequency band power
fMRI	Functional magnetic resonance imaging
ICT	Information and communication technology
ISI	Interstimulus interval
ITU	International Telecommunication Union
ITU-T	International Telecommunication Union—Telecommunication Standardization Sector
LDA	Linear discriminant analysis
MEG	Magnetoencephalography
MMN	Mismatch negativity
MOS	Mean opinion score
MUSHRA	Multi stimulus test with hidden reference and anchor
NS	Not significant
OP	Opinion test
POLQA	Perceptual objective listening quality assessment
QoE	Quality of experience

ROC	Receiver operating characteristic
SAM	Self-assessment manikin
SNR	Signal-to-noise ratio
SVM	Support vector machines

Chapter 1
Introduction

Research and development in the area of information and communication technology (ICT) has been growing over the last two decades.[1] Not only this part of the industry has grown, it is also embedded in research and innovation funding initiatives of such as the Horizon 2020 program of the European Union (EU). This development has been accompanied by growing numbers of available ICT services, and consequently, by an increased number of customers. One of the most visible results of this development is that the percentage of customers with a mobile phone subscription has increased from 29 to 132 per 100 inhabitants within the decade from 1999 to 2009 [6].

With a subscription, a wide range of possible functions becomes available for customers. On the one hand, these devices enable customers to use standard functions such as making calls and texting, and on the other hand, extended functions like internet access and media steaming become available. A growing number of services provided in the form of apps are available for users and easy access is provided, e.g., from Apple through the AppStore. With this increased number of users and services, telecommunication companies have to offer (i) good performance of the offered services, and at the same time they have to (ii) deal with the quickly growing data volume that has to be handled and shared by the provided services. With this one very well known attempt of the industry to meet both of these requirements are codecs, more specifically, speech and video codecs, e.g., [7] for the wideband encoding of speech and [8] for video encoding, respectively. The main idea behind media encoding is to reduce the amount of data that needs to be transmitted while keeping the perceived quality of the transmitted media as high as possible. Services that deliver information to the user can be categorized according to the possibility of determining the performance of the system [9]. For services or applications that include

[1] Parts of this chapter have been previously published; text fragments and figures have been based on Antons et al. [1], reprinted, with permission, from Antons et al. [1], Antons et al. [2], reprinted, with permission, from Antons et al. [2], Antons et al. [3], reprinted, with permission, from Antons et al. [3], Antons et al. [4], reprinted, with permission, from Antons et al. [4], and Antons et al. [5], with kind permission from Springer Science+Business Media: Antons et al. [5].

J.-N. Antons, *Neural Correlates of Quality Perception for Complex Speech Signals*, T-Labs Series in Telecommunication Services, DOI 10.1007/978-3-319-15521-0_1

human-machine interaction, *usability* can be utilized as a measure of performance. According to the ISO definition 9241 Part 11 (1999), usability includes not only the classical performance measures "efficiency" and "effectiveness", but also the satisfaction of users with the given interface or service in general [10].

For interactive communication services such as speech or video conferencing, performance can be assessed in terms of perceived conversation quality. In the event that media information is delivered to the user (e.g., listening to auditory signals or speech), the transmission quality and the quality of the media itself impact the perceived quality. In general, it is evident that the quality as perceived by the user, and therefore maybe the acceptance of the service, are of prime importance. One example which substantiates this direct impact was presented by Chen et al. (2006), who demonstrated that Skype calls lasted longer and more calls were initiated if the quality of the transmitted media was high [11].

The concept of *quality* is defined by Jekosch as "judgment of the perceived composition of an entity with respect to its desired composition" [12]. The term "quality" is also already widely known from the perspective of the manufacturer or service provider as the concept "Quality of Service" (*QoS*), which has been defined as: "The totality of characteristics of a telecommunications service that bear on its ability to satisfy stated and implied needs of the user of the service" [13].

One of main focuses of this book is to establish a test set-up which can be used as a complementary measure during quality tests, whereby the focus is placed on easily controllable quality variations with an impact on user perception, and therefore, on *Quality of Experience* (*QoE*) as well. For a detailed overview on factors governing QoS see [14]. An updated and most recent definition of QoE with a focus on user aspects has been formulated by Raake and Egger (2014): "Quality of Experience is the degree of delight or annoyance of a person whose experiencing involves an application, service, or system. It results from the persons evaluation of the fulfillment of his or her expectations and needs with respect to the utility and/or enjoyment in the light of the persons context, personality and current state" [15].

The definition of QoE is much broader than the QoS definition, as there is, e.g., no direct mention of telecommunication services. Whereas the definition of QoE also includes hedonic components such as "enjoyment", the QoS definition focuses more on the utilization such as "implied needs" [16].

Aspects of QoE are mainly assessed by asking participants to express their opinion on the presented media. For telecommunication systems, subjective assessment methods are standardized by the ITU in, e.g. , ITU Recommendations [17] (for audio quality), [18] (for video quality), and [19] (for speech quality). In telecommunication research, speech, audio, and audiovisual quality are typically assessed with behavioral tests in which participants provide a rating corresponding to their impression (a detailed description of subjective methods for speech quality assessment is given in Sect. 1.1). These methods represent state-of-the art techniques to evaluate the quality of a system or service in the area of speech quality assessment. Nevertheless, these methods are strictly limited to conscious evaluation and are therefore only accessible after the introspection of listeners.

No single method used for QoE assessment can capture all relevant factors simultaneously, such as the resulting overall quality rating and the impact on the cognitive state of listeners. If two measurement techniques are used in sequence, an effect on the final result can be produced due to sequence effects. For example, it was shown that if quality and emotional ratings are requested in sequence, the order of the questioned items had a significant effect on the final result, i.e., the quality was rated lower if it had not been asked as the first item [20]. Therefore, it is important to consider potential sequence effects when applying several methods.

Neurophysiological data provides one promising approach and can complement subjective ratings as a comprehensive and non-intrusive measure, potentially revealing neuronal differences in quality processing below the threshold of conscious perception that might affect the long-term satisfaction of a user. In general, only a fraction of all perceptual processing enters consciousness and is as such available to the introspection required for self-reporting [21]. Nonetheless, non-conscious processing steps are also accompanied by neuronal changes, and thus, physiological measures may provide insight into how these processes eventually lead to a given rating. In addition to this, it is possible to record on a continuous basis and obtain data closer to the actual auditory event (with respect to time). A detailed introduction to the physiological methods used in this book is given in Sect. 1.3. In general, two different analytical techniques can be considered: (1) changes in the *frequency power of electroencephalogram bands* in relation to the presented stimuli (see Sect. 1.3.1) and (2) *time-locked responses* (see Sect. 1.3.4). In psychological research, a great amount of behavioral experiments have been conducted in order to to assess how humans perceive external stimuli [22]. Following significant developments in imaging technology, research also extended into fields concerned with how perceptual stimuli of different modalities are processed by the brain [23]. In recent studies, these imaging techniques have been applied to stimuli that are of interest for QoE research. Miettinen et al. (2010), for instance, showed a significant increase in the measured amplitudes of magnetoencephalography (MEG) signals for distorted stimuli [24].

This book will answer the question of how the combination of a standard subjective test and physiological measurements can help to improve our understanding of the ways in which media quality can affect subjective experience. Potentially, high satisfaction on the side of the customers can be ensured, while service providers can gain a good marketing argument based on their delivery of high-quality products. As regards the latter effect, it is clear that high-quality transmissions are used for advertising purposes. The connection between high quality and customer acquisition has not yet been successfully analyzed. In the following paragraphs, an overview will be presented about how speech quality can be subjectively assessed (Sects. 1.1, and 1.2). This will be followed by an introduction of physiological measures which can be used for assessing quality-related brain signals (Sect. 1.3).

1.1 Quality and Quality of Experience

Complex speech signals are transmitted in order to be consumed by human users. The acceptance and experience of quality are the main success factors of services like telephoning or streaming services [25]. Quality research is therefore one key area for enabling high-quality services. According to [26], quality is the result of a perception and a judgement process. Therefore, it is only assessable after a subjective test has been carried out. During this process, perceived characteristics are compared to individual expectations. Furthermore, Möller postulates that quality is only measurable by testing participants, and that quality is an event which is subject to the situational influences under which it is perceived or judged.

QoE was defined in [15] as "the degree of delight or annoyance of a person whose experiencing involves an application, service, or system. It results from the persons evaluation of the fulfillment of his or her expectations and needs with respect to the utility and/or enjoyment in the light of the persons context, personality, and current state".

In the following section, an introduction and definition of the term "quality" as well as its relation to possible physiological parameters will be given. After this, subjective techniques for the assessment of auditory speech stimuli will be presented (Sect. 1.2). The chapter will end with an overview on the physiological measurements used (i.e., frequency band analysis in Sect. 1.3.1, and event-related potentials in Sect. 1.3.4).

According to the considerations of Jekosch [12], speech quality assessment consists of a three-step process: perception, judgment, and description. The first step comprises the reception of the "physical event" (e.g., the speech sound wave reaching the human ear) and its transformation into an "auditory (hearing) event". The judgment process, in turn, is responsible for deriving features (e.g. loudness or coloration) from the perceived event and comparing them to internal reference criteria defining how speech of good and bad quality should sound. This internal reference is unique to each listener and may be influenced by numerous factors, such as the users expectations, experience, motivation, affective state, and ambient factors, to name just a few. Lastly, the final description process involves the pooling of these judgments into a final quality rating.

A detailed definition of quality which emphasizes the importance of the fact that all features of a sound must be recognizable and namable was given by Blauert [27], which stresses the subjective nature of quality.

Factors that can influence the Quality of Experience can be assigned to three categories: (1) human, (2) system, and (3) context [28].

The *human influencing factors* are defined as properties or characteristics of a human user, which can be an invariant factor, e.g., demographic information about potential system users, or variant user characteristics such as mental condition. Such factors are very important for this book, as physiological techniques can potentially record the mental state of participants. In a strict sense, such factors influence the Quality of Experience of users and in the following experiments, the cognitive state of users will be measured as indicators of changes caused by the presented stimulus material.

Factors with a capacity for a systemic influence (*system influence factors*) are all properties that constitute the technically produced quality of a product. In the experiments conducted here, factors with a capacity for systemic influence have been used as independent variables. Intended influences of systemic parameters as well as subjective and physiological responses are of interest.

Contextual factors (*context influence factors*) which influence the QoE include all situational factors that can be used to describe the environment of users. Due to the fact that all experiments in this book were performed in a laboratory, such influences were kept to a minimum and stabilized.

The term "feature" is used as a dimension of the perceptual space in which users rate their perception of QoE influenced by the aforementioned factors [29]. A quality feature is defined by [12] as: "A perceivable, recognized, and namable characteristic of the individuals experience of a service which contributes to its quality". Quality-relevant perceptual features of speech services are: discontinuity, noisiness, and coloration [30].

In this book, quality will be used and measured as the final integrational factor in the subjective rating process; a splitting off and separate discussion of speech features and possible subscales will not be undertaken.

Possible *temporal developments of QoE* will also not be considered. When listeners are exposed to speech in varying quality, the temporal position of a degradation does not has the same impact at all times [31].

In order to mitigate the negative effects of variability between participants, subjective listening tests such as the Mean Opinion Score (MOS) test [19] average all listener quality ratings.

According to [32], physiological processes within the participant—which are not observable during a standard subjective experiment—can be represented in the simplest cases as a "black box" with three ports, one input (the sound event) and two outputs (the auditory event and the description).

The updated definition of Quality of Experience (QoE) from Raake and Egger (2014) can be visualized as a simplified structure [15].

In Fig. 1.1, a simplified version of the *quality formation process*—which is based on [12, 14, 15, 32]—has been visualized. Note that *boxes* outside of the participant represent input and output including the incoming signal (*physical event*) and contextual parameters. Marked in red are the parameters used in this book to influence the input (experimental manipulations and physical event/stimuli) and the measuring criteria *event-related potentials* (*ERP*), *alpha and theta frequency band power* (*FBP*), and subjective ratings (*opinion test*). After the signal has been processed at the sensory level by the perceptual system, the internal representation and perceived quality are compared. This comparison results after encoding and quantitative factors have been assigned to a quality rating. Throughout this process, all stages can be influenced by the cognitive state of the listeners. Of course, the latter process can also be reversed and the perceptual process can influence the cognitive state. On the one hand, this book will utilize standard subjective techniques to assess the speech quality rating of participants by using opinion tests. The data used in this approach will be the result of the quality formation process, i.e., the rating itself. On the other

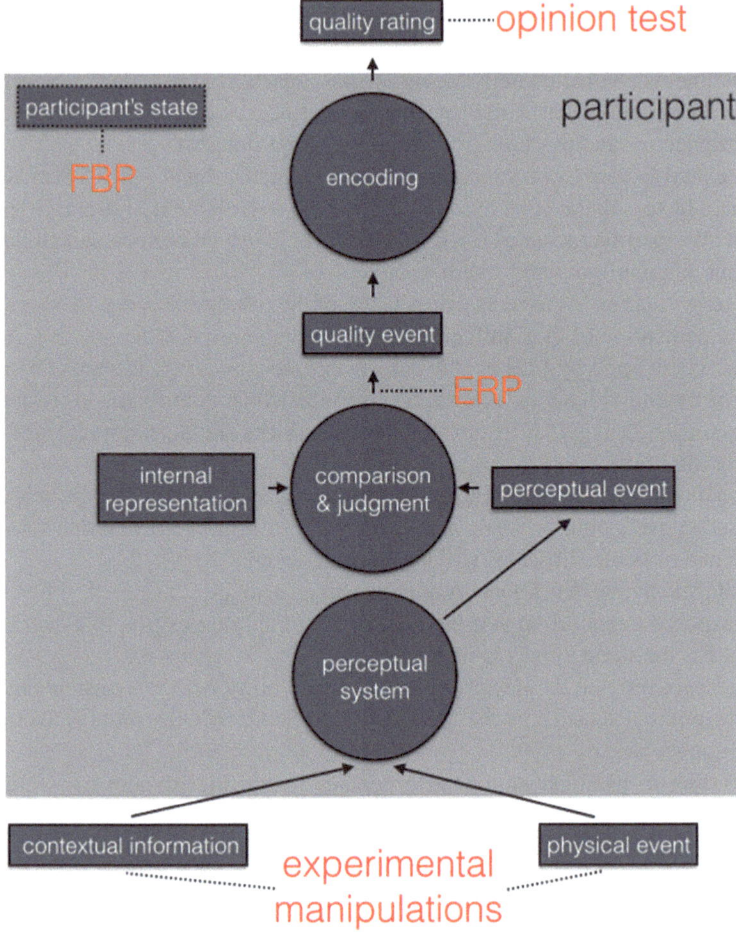

Fig. 1.1 Simplified version of the quality formation process (based on [12, 14, 15, 32]). *Circles* represent perceptual processes, *boxes* represent storage for different types of representation. Note that *boxes* outside of the participant represent input and output including the incoming signal. The comparison of internal representation and perceived event will effect a quality rating after encoding. Marked in *red* are the parameters used in this book to influence the input (experimental manipulations; contextual information and physical event/stimuli) and the measuring criteria *event-related potentials* (*ERP*), alpha and theta *frequency band power* (*FBP*) and subjective ratings (*opinion test*). The (cognitive) state of the participant can influence all stages of the formation process and the process itself can influence the participant's state. Note that this picture does not include either the detailed anticipatory process or the detailed comparison and judgment process (see [15] for a more detailed model). For a detailed description see text in Sect. 1.1

hand, physiological measuring techniques will be used to infer the neuronal response towards stimuli of varying quality. Firstly, the impact of stimulation with speech of varying quality and its effect on the cognitive state of listeners will be addressed, which can be measured on the basis of the frequency band power (FBP, alpha frequency band power) generated by physiological activity and recorded by an electroencephalogram (EEG) (see Sect. 1.3.1); and secondly, the neuronal response to the comparison between internal representation and perceived quality will be addressed, which results in varying P300 responses (see Sect. 1.3.4.5). The positioning of the ERP parameter within the model—between the comparison and the quality event—is based on the fact that ERP components can be conscious and non-conscious (for an detailed overview on ERP components see Sect. 1.3.4).

1.2 Quality Assessment Methods for Speech Stimuli

The standard methods for the speech quality assessment of short stimuli can be divided into two classes: those with and without reference. Methods without reference (e.g., *Absolute Category Rating* (*ACR*)) result in a *Mean Opinion Score* (*MOS*) based on judgments of the test stimulus alone.[2] The MOS is a numerical value commonly expressed on a scale from 5 (excellent) to 1 (bad). For the ratings of high-quality samples, those methods are suitable which have a reference stimulus. During such tests, participants must rate the quality of an experimental stimulus compared to the quality of a reference sample. The *Comparison Category Rating* (*CCR*) and the *Degradation Category Rating* (*DCR*) are examples of such reference-based methods used in speech quality assessment [19] and will be addressed in a later part of this chapter; similar paradigms are available for visual and audiovisual stimuli.

For stimuli of higher quality, it is also possible to assess multiple stimuli simultaneously. This can be done, for example, by using the *MUlti Stimulus test with Hidden Reference and Anchor* (*MUSHRA*) methods in accordance with ITU-R Recommendation BS.1534-2 [33]. Utilizing this method, listeners must rate multiple stimuli on a *continuous quality scale* (*CQS*). The SQC is a graphic scale divided into five equal intervals with the predetermined adjectives: excellent, good, fair, poor, and bad (see Fig. 2.2 in Chap. 2). The rating results in values ranging from 100 (excellent) to 0 (bad). The CQS is also used in ITU-R Recommendation BT.500 for the evaluation of pictorial quality [17].

As these subjective techniques are often very costly and time-consuming, instrumental techniques have been developed. These instrumental techniques consist of computer programs which were designed to automatically estimate the perceived quality of transmitted media, in particular, of speech. Two quite well established instrumental models are the *E-Model* [34] and *POLQA* [35], which will be utilized in Chap. 6.

The QoE definition also includes the emotional impact on the listener: "Quality of Experience is the degree of delight or annoyance of a person whose experiencing involves an application, service, or system". Delight and annoyance are strongly

[2] When presenting several test stimuli in a series, contextual factors may cause the judgment to not only be influenced by the test stimulus alone, but also by other stimuli which form part of the test set.

related to the emotional response a user has towards a certain product. According to [36], an emotional response can be evoked by the content of a media stimulus (e.g., content of a audiovisual stimulus), or the quality relevant to experimental parameters (e.g., video resolution). In this book, the latter factor will be focused on. The experienced affect can be assessed by the *self-assessment mannikin* (*SAM*), which is a pictorial scale depicting a simple cartoon figure. The expression of the cartoon varies in three dimensions: valence (unhappy to happy), arousal (sleepy to exited), and dominance (being controlled to being in control). The SAM will be used for assessing the emotional state of participants in Chap. 4.

Although these techniques provide good estimations of quality judgments, they do not provide insights into how the participants form these judgments. The antecedent processes are the key elements to fully understanding how ratings are formed and may also influence the cognitive state, e.g., vigilance.[3] Electrophysiological measuring techniques utilizing electroencephalographic data (EEG) can help improve our understanding of the impact caused by media quality on subjective experience (see overview in the Sect. 1.3).

Participant-dependent experiments require living participants who are able to judge the presented stimulus material. The most common way of conducting subjective experiments will be explained in the following sections based on [26]. User tests can be passive or active. Interactive scenarios are used for testing when it is important to imitate actual telephoning situations. In passive scenarios, users are exclusively listening to the stimuli presented to them and asked to rate their quality. Quality rating can be undertaken in various ways and many different quality scales can be used [26] and [38]. If the overall quality rating without references is of interest, the *Absolute Category Rating* (*ACR*) method is used. If small differences between two signals are from interest the *Comparison Category Rating* (*CCR*) or the *Degradation Category Rating* (*DCR*) techniques can be applied. The latter rating uses an annoyance scale (from "degradation is inaudible" (5) to "degradation is very annoying" (1)) and the CCR uses a scale on which the amount by which the second (quality-manipulated) sample is degraded relative to the unprocessed first sample (reference); the scale uses a range from "much better" (3) to "much worse" (−3). Both ratings use a reference prior to stimulus which has to be evaluated, and both are suitable for evaluating high quality speech [19].

For benchmarks of two, e.g., speech transmitting systems, *Pair Comparison* (*PC*) can be used.

In most cases, a 5-point scale is recommended which ranges from 1 to 5 (5 = excellent, 4 = good, 3 = fair, 2 = poor, 1 = unsatisfactory) [39]. Quality scores are collected and commonly averaged across all test participants in order to obtain the *Mean Opinion Score* (*MOS*) [26].

A *psychometric function* relates the performance of participants, e.g., in a detection task, to an independent variable such as a stimulus parameter (varying quality) [40]. In the context of this book, psychometric functions are used to display the performance

[3] Vigilance is defined by Parasuraman (2000) as "... sustained attention—ensures that goals are maintained over time" (p. 7) [37].

in detecting that a stimulus has been degraded—rated in an absolute two-choice forced task as a function of degradation intensity. Performance expressed as the ratio of degraded detected stimuli to the overall number of presented stimuli is used in this context in Chaps. 2 and 3.

The perception of transmitted speech: Speech which is transmitted though a communication system is modified in terms of a comparison between the physical signal emitted by the mouth of the actual speaker and the physical signal generated from the technical equipment at the receiver side of a communication [25], and [30]. Following [30] these modifications can be classified as the following "perceptive factors":

- Loudness,
- articulation,
- perception of the effects of bandwidth and linear frequency distortion,
- perception of sidetone,
- perception of echo,
- perception of circuit noise,
- effects of environmental noise and binaural hearing, and
- effects of delay.

Quality assessment: Subjective tests that discriminate between the quality of speech systems are termed "speech quality tests" [39]. In contrast to intelligibility tests, the intelligibility in speech quality tests is in most cases high, meaning that the listener is able to understand the content under all conditions. In an generalized sense, any test requested subjective judgment can be called *opinion tests*, as the main task of participants is to verbalize their opinions. For general considerations on test set-ups concerning QoE, see [26] and [38].

The scaling of ratings is usually carried out in two ways, either direct scaling or indirect scaling [26]. Direct scaling has the advantage that the listener must rate on a predetermined scale, meaning that the attribution of a physical event (speech sound) to a perceptual event scale is required (e.g., the quality is poor). For indirect measurement, the listener must set a certain value or detect a degradation without designating an external reference point (e.g., the second stimulus is of lower quality).

The level of measurement can be divided in three main classes: the nominal, ordinal or interval, and the ratio scale [26]. At the nominal level, a differentiation is only carried out based on the name, whereas using a scale at the ordinal level allows the data to be sorted by rank order (one value is greater than another). The interval scale is an ordinal scale but has an equidistant distribution of values (one value is twice as much than another). The ratio scale exceeds all other levels of measurement, as all statistical measures are allowed due to the existing zero value [26] and [38].

1.3 Electrophysiology and Electroencephalogram

In addition to the well-known and approved approaches to quality measurement, the *electroencephalogram* (*EEG*) has proven to be a valuable tool for quality research in the auditory and visual domains, which can provide additional information about underlying processes [41, 42, 43, 44, 45].

The electroencephalogram (EEG) measures voltage variation caused by neuronal activity in the brain by attaching electrodes to the scalp of a participant. Since its discovery by Berger in 1929, it has become a widely used method for investigating the physiological correlates of perceptual and attentional processes [23, 46]. This measuring device has a rather limited spatial resolution—due to the fact that the brain is a wet conductor, the signals recorded by one electrode are mixture of all existent sources, but with an excellent temporal resolution that has a precision of milliseconds.

Scalp-recorded EEG oscillations can presumably be referenced to a large number of synchronously activated pyramidal cortical neurons [47]. This is due to the summation of synchronous excitatory and inhibitory post-synaptic potentials. When high frequencies are generated, corticocortical and thalamocortical interactions have been postulated. In addition to the large-scale synchronization of many thousands of neurons, localized synchronization also seems to be important for EEG generation (see [47] for a detailed description of EEG generation).

The corresponding data can mainly be analyzed in two different ways: on the one hand, by taking a closer look at the spectrogram of spontaneous activity, and on the other hand, by analyzing so-called *Event-Related Potentials* (*ERP*) which are a time-locked reaction to an external stimulus measured as a change in voltage [48]. The latter approach can be used to analyze cortical potentials as well as voltage differences generated in the brain stem.

ERPs are of special interest to the neuroscientist and will be described in more detail in the following sections. Fabiani et al. (2007) consider ERPs to be: "... one of the main tools available to cognitive neuroscientists" (p. 110) [49]. In the context of EEG-based *Brain Computer Interfaces* (*BCI*) [50], machine learning methods play a crucial role in extracting relevant information from high-dimensional data [51] and [52]. In recent years, there has been an increasing interest in non-medical applications of BCI technologies [53] and [54].

This book will focus on cortical brain activity and just briefly mention brain stem measurements, as QoE research at the brain stem level is not yet widely available for many degradation classes and the standard length of stimuli. In addition to this, the signal-to-noise ratio for these kinds of measurements is so low that the stimuli must be presented numerous times, and due to the duration of the resulting experimental set-up only few stimulus classes can be presented per experiment.

Besides the relevant information—brain activity—much unwanted information is recorded as well, e.g., voltage changes due to eye movement body movement, and other irrelevant signal sources. Due to the high noise in the signal, it is important to create strictly controlled experimental set-ups. Clinical research guidelines for experimental designs already exist, and in this book important implications for research in the domain of Quality of Experience will be outlined based on [46].

Descriptions will be provided about how to analyze continuous and evoked EEG data and how the two ways of analyzing the EEG signal are carried out; and how these techniques have already been used for studies concerning Quality of Experience. Recently, new *low-cost EEG devices* have appeared on the market, such as the Emotiv-EPOC[4] and NeuroSky MindWave[5] headsets. Although these consumer products are comparatively inexpensive, the data quality, i.e., the precision and noise content of these products, are much worse compared to the devices used in clinical applications. However, these products have proved their ability to record useful information in the context of QoE related research.

Moldovan et al. used the criteria provided by the Emotiv EPOC System to infer the level of frustration of human observers caused by the quality of the audiovisual excerpt presented to them. This level was ascertained by using a metric predefined by the headset manufacturer. In the presented videos in which different levels of quality were used, scientists manipulated the bit rate, frame rate, and resolution of the video clips presented to test participants [55].

Perez et al. (2013) used the NeuroSky MindWave headset to measure brain activity and utilized the recorded data to classify their test trials into high-quality and low-quality pictures [56].

1.3.1 EEG Frequency Band Power

One important question is how phasic, i.e., short-term changes might eventually lead to tonic, i.e., long-term effects on the state of the listener. To address this question, an approach can be taken to assess the condition of the user in terms of her/his *cognitive state* when being confronted with, e.g., speech stimuli with different audio bit rates, by analyzing the spectral components of *electroencephalographic* (*EEG*) signals.

Well-known test cases for accessing the condition of a user and estimating his/her cognitive state/fatigue by means of spectral EEG components are car drivers. In general, there are five different frequency ranges ascribed to specific states of the brain: *delta band* (1–4 Hz), *theta band* (4–8 Hz), *alpha band* (8–13 Hz), *beta band* (13–30 Hz), and the *gamma band* (36–44 Hz) [47]. The *delta band* is present during the deep sleep of participants, the *theta band* during light sleep and is also an indicator for decreased alertness. Activity in the *alpha band* is related to relaxed wakefulness with eyes closed and a decrease in alertness. *Beta and gamma band* are ascribed to high arousal and focused attention [48].

Analyzing the power in the aforementioned frequency bands is widely done for assessing the *cognitive state* of car drivers. Lal et al., for example, showed that

[4] http://www.emotiv.com/.

[5] http://www.neurosky.com/.

fatigued drivers had elevated activity in the delta and theta bands [57]. Correlations between the weighted combinations of different frequency band powers with subjective fatigue ratings were shown in [58].

Another reason to use frequency bands is the ability to estimate the emotional state of participants. For this purpose, alpha values from frontal electrodes are extracted. The *asymmetry index* is one way to obtain this information. This index shows that higher values in the asymmetry index are the result of higher left frontal lobe activity,which is usually due to negative emotional processing [59].

The *delta band* is present when participants are asleep and reflects low-frequency oscillations (1–4 Hz) [47]. With the exception of pathological cases, delta band activity is not predominant in the normal waking consciousness of human beings, as it seems to be inhibitory in essence.

The *theta band* is known to be present during light sleep but also an indicator for decreased alertness (drowsiness) and impaired information processing. A connection between a predominant theta band and the gating process of information processing has been postulated [47]. This band could presumably be an indication—if measured as a predominant factor—of not fully normal functionable information processing. This could also be of interest in connection with transmitted speech as well. In addition to the established association of the theta band to decreased alertness, it is also known to be increased during high workload conditions [60].

Increased activity in the *alpha band* is related to relaxed wakefulness with eyes closed or a decrease in alertness. This band is known to predominate during the fatigued state of participants, e.g., while driving a car. In the later stages of sleep, the alpha band is suppressed. In this book, participants in a sleeping state have not been discussed, therefore, the increase of the power in the band has been interpreted as a reduced cognitive state (drowsiness).

The higher-level bands *beta and gamma* are ascribed to arousal and focused attention [48].

Technique for measuring frequency band power will be applied in Chaps. 5 and 6.

1.3.1.1 Findings Related to QoE

In the context of fatigued drivers, Lal et al. showed a high reproducibility for delta and theta bands as indicators for fatigue [57, 61, 62]. For an overview of alpha frequency band power as a measure of mental workload see [63]. The author discusses changes in the alpha frequency band power which can be used to assess the state of the drivers. In a recent experiment from Punsawad et al. (2011), a correlation between the weighted combinations of different frequency band powers with subjective fatigue ratings was shown [58]. Most interesting for possible applications is the use of a single electrode, which can help to minimize preparation time and create a more comfortable situation for participants.

As it is possible to utilize more natural stimuli with longer stimulus lengths, the effect of longer lasting media stimuli (>10 min) on recipients can be analyzed. In another study, extracted alpha values from frontal electrodes were also used in order to assess the emotional state of test participants. It could be shown that higher-level left frontal rather than right frontal lobe activity was recorded, which therefore indicated an increased asymmetry index. In this case, the response of these participants after their exposure to low-quality stimuli indicated an emotionally negative processing of these stimuli, in contrast to higher-quality stimuli; respective correlations to subjective scores were also presented [64]. In order to determine the asymmetry index, the relationship between left frontal and right frontal lobe activity must be calculated; this is carried out by using the corresponding alpha proportions: $(ln(alpha_{right}) - ln(alpha_{left}))$ as proposed by Coan and Allen [59]. Even though the asymmetry index could be potentially useful to assess, e.g., satisfaction with a service or product, it will not be employed in this book.

The results of the studies discussed here indicate a correlation between the derived parameters—brain patterns related to attention/fatigue (arousal/drowsiness) and emotions (valence)—subjective QoE parameters such as subjective, emotional self-assessment, and quality ratings.

1.3.1.2 Data Acquisition and Analysis

As a continuous signal is not related to one brief and isolated event, this method is suitable for stimuli of longer duration. Usually the intervals for analysis are between 5 and 10 min long and are set in relation to a baseline interval from 2 to 5 min, resulting in a baseline-corrected power value.

For this analytical method, a small set of electrodes is frequently used; approximately 8 electrodes are sufficient and should be distributed at occipital/parietal scalp locations for attention and fatigue studies, and frontally for asymmetry index studies in accordance with the 10–20 *system* which assigns electrode positions based on the ratio distance from a central point on the scalp [65].

Most interesting for the possible employment in practical applications is the use of a single electrode, minimizing preparation time and making the application more comfortable for test participants. Of course, less electrodes result in less information as a result of reduced spatial electrode distribution, which also limits the possibility of dealing with noise (e.g., independent component analysis) and dipole source estimations due to the reduced spatial information.

1.3.2 Working Memory, Vigilance, and Cognitive State

For every presented speech signal, it is important to enable the listener to remain concentrated over time, whether following a conservation over the telephone is involved, or listening to an audiobook in the attempt to follow its plot. Therefore, *vigilance* is one important factor for enabling high-quality speech presentations to listeners. Vigilance is defined by Parasuraman (2000) as "... sustained attention—ensures that goals are maintained over time" (p. 7) [37]. Warm et al. (2000) describe vigilance as the ability to maintain concentrated attention over prolonged periods of time.

When talking about vigilance, it is important to mention *mental workload* as well. The assessment of human mental workload is of especially high importance when considering cognitive engineering applications [66]. For tasks with a high risk of, e.g., injury, it is important that the workload is not too high or too low. An assessment of mental workload can be done by analyzing behavioral data such as accuracy and speed of responses (reaction times). In addition to there standard measurements of behavior, physiological parameters can also be indicative of changes in mental workload. Thus, a high mental workload condition will result in a reduced P300 peak amplitude when recorded in a dual task situation.

In addition to the use of EEG frequency band power for estimating the cognitive state—the *vigilance level*—of participants, *reaction times* can be used to measure the current cognitive state of participants.[6] Of course, task performance measurements expressed as reaction times are also indicators of the difficulty of the task itself as well as the ability of the participant, but if the task difficulty is kept constant, limited cognitive resources can be measured by decreased task performance, i.e., slower reaction times.

The fact that the measurement of reaction times can be used for the assessment of the cognitive state of humans has been demonstrated, e.g., in [67]. In this study, it was shown that the average 80th percentile (separately calculated for each participant) was measured as the value for the slow reactions of participants, and the mean of all reaction times below the 20th percentile represented the fast reactions of participants. Both the mean fast and slow reaction times were indicative of the cognitive state, i.e., sleepiness of the drivers. In general, reaction times can be analyzed as a measure of user performance and their cognitive state. Of course, reaction times can vary for several reasons and experimental conditions must be randomized so that variance non-attributable to experimental variations is excluded by averaging.

1.3.3 EEG Experiment

There are several important factors—such as the number of electrodes or the sampling frequency—that have to be considered when planing a physiological experiment in combination with a QoE-related research question. A good overview on important parameters for physiological experiments has been provided in [47] and [68], in the next paragraphs only a summary of the most important facts will be given.

[6] Reaction times can be measured by determining the needed time to complete a designated task. The given task can either be the main or secondary task.

1.3.3.1 Electrodes

The *electrode location* and the *number of electrodes* is strongly dependent on the objectives/components of the measurement. First of all, it is clear that the electrode it-self should not attenuate the signal within the frequency band of interest (usually between 1 and 70 Hz).

For the placement of electrodes, placement arrangement systems are available such as the 10–20 *system* [65], which aim to place the electrodes homogeneously on the scalp and have comparable measurement positions between participants. A distance of approximately 2 cm between electrodes is necessary in order to prevent distortions of scalp potential distribution [47].

The number of electrodes used can be varied in standard set-ups from 1 up to 256 electrodes. A higher number of electrodes can be useful, if electrode interpolation or an advanced cleaning of EEG data is done.

Types of electrodes can be divided into two classes: (1) wet and (2) dry. The latter class is quite new and has just recently become commercially available, but has the advantage that the small metal pins have direct contact with the scalp and no abrasion fluid is necessary. In addition to this, the time needed to set up the equipment is shorter, but participants can experience uncomfortable pressure or headaches due to the required pressure on the metal pins. Wet electrodes have the advantage that once applied they can be used for longer experiments and are not so sensitive to motion compared to dry electrodes.

The reference electrode—as the voltage difference measured by the EEG is always a difference in potential between two electrodes—is one of the more important electrodes. Therefore, placing is also crucial, and should—if possible—be done in an electrically inactive position. As the human body itself is a wet conductor, this is not absolutely possible. In many experiments, places with low electrical activity were selected, such as the nose or the mastoids (bone structures behind the ear channel filled with air).

1.3.3.2 Recording

During the recording itself—due to the *Nyquist Theorem*—the sampling rate (measuring points per second) should be at least twice as much as the highest frequency of interest. If this condition is not met, frequencies higher than half of the sampling rate should be removed [47]. As artifacts during most measures can be noise of the power line (50/60 Hz), which can be filtered out using a notch filter.

1.3.3.3 Analysis

The analysis of EEG data can usually done by (1) analyzing the spectral domain, (2) doing a time-frequency analysis, (3) a coherency analysis, and analyzing the ERPs [47] and [49].

Analyses in the spectral domain are usually carried out using the *Fast Fourier Transformation* (*FFT*) to estimate the power of various frequencies in the signal. With these techniques, it can be estimated what frequency band is predominant (for an explanation of different frequency bands of EEG data see Sect. 1.3.1), and which cognitive state is associated with it.

Asymmetry metrics—e.g., showing the difference in activation for the two brain hemispheres—can be used to indicate whether two brain regions are differentially involved in a cognitive task or affective process.

Analyses in the time-frequency domain are called "spectra" and can be calculated by using a *short-time Fourier Transformation* or doing a wavelet. Either way, it can be beneficial to be able to analyze non-stationary phenomenon such as transient changes in the frequency domain.

Coherency analysis can be carried out to determine dynamic interactions between EEG signals—recorded at different scalp locations—and can be calculated similar to a correlation by dividing the cross-spectrum between two time series by the root of the two spectra [47].

The *analysis of ERPs*—the phase-locked information evoked after stimulus presentation—will be explained in Sect. 1.3.4.

In this book, EEG data will be predominantly analyzed by inspecting the spectral domain and ERPs.

1.3.4 Event-Related Potentials

The second way of analyzing EEG data is to check for event-locked information after recording continuous EEG data (Sect. 1.3).

1.3.4.1 Stimulus Presentation

The stimulus presentation for many ERP experiments carried out by using an *oddball task* [69]. During this task, stimuli are presented in a sequence. Between stimuli there is either a fixed or jittered *interstimulus interval* (*ISI*). The next stimulus will in most of the cases start automatically, without the necessarity of user input.[7] In many experiments, two stimuli are used, one which presented more frequently—the *standard*—and one presented irregularly—the *deviant/target* (see Fig. 1.2).

[7] In most ERP experiments, participants must react to the presented stimuli, which can include a reaction after the presentation of one stimulus class or the forced choice to sort the last stimulus into a given category.

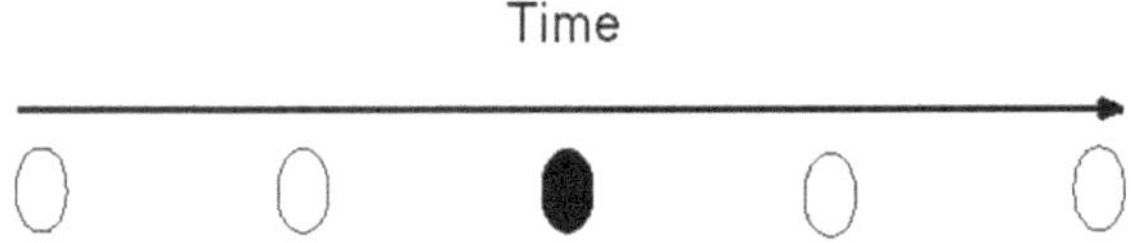

Fig. 1.2 Schematic representation of the *oddball task* paradigm (based on [70]). *White circle* represents a standard and *black circle* represents a deviating target stimulus. In terms of a quality experiment example, stimulus classes are stimulus with high quality as a *standard* and a stimulus in lower quality as *deviant*

1.3.4.2 Measuring Procedures

Measuring procedures for ERP components relevant to measurement and how to carry out their analysis has been described in detail by, e.g., [46, 69, 71]. Here, an overview of the most important facts will be given.

As already mentioned in Sect. 1.3, the signal measured during an EEG recording is the difference potential between two electrodes. As the intended measurable frequency band is between approximately 1 and 70 Hz, frequencies within that band should not be attenuated [47].

For improving the signal-to-noise ratio, several techniques have been proposed by [49]. Firstly, frequency ranges irrelevant to test objectives can be attenuated—using analog or digital filters—as most endogenous ERP components are between approximately 0.5–20 Hz. Secondly, multiple recordings of the signal during presentations of the same stimulus can be averaged. As the recorded signal is a mixture of the ERP itself and non-time-locked activity (noise), the latter signal will be attenuated by averaging.

The naming of ERP components is carried out using the polarity and timing of the peak. A positive peak around 300 ms after stimulus onset is therefore named "P3" (third positive peak) or "P300" (positive peak approximately 300 ms after stimulus onset). The naming procedure is consistent for other components as well.

The number of electrodes is strongly dependent on the research objective. In a classical BCI setting, a high number of electrodes is desirable [54]. In a setting with more practical application and when it is evident which strong ERP components are targets of the measurement, set-ups with a few electrodes can be sufficient [58].

One important implication demonstrated by [47] is that as soon as source reconstruction techniques are used—a mathematical technique reconstructing the dipolar source of neuronal activity—high-density recordings with an equal spatial sampling become necessary.

The equal spatial sampling of the signal is ensured by placing electrodes at approximately the same relative sites on the scalp [47]. One of the most common methods to ensure a standardized—i.e., between different participants comparable—system of electrode placement is the extended 10–20 system [65]. The electrodes are placed at sides 10 and 20 % from nasion (which is the crossing point of the frontal bone and the two nasal bones of the human skull, approximately between the eyes), inion (most prominent protrusion of the lower rear bone of the skull), left and right *mastoid* (bone behind the outer ear channel).

A basic parameter of every EEG experiment is to determine how many *trials* are necessary in order to reduce the *signal-to-noise ratio* (*SNR*). The signal-to-noise ratio can be influenced by the number of trials, due to the fact that activity unrelated to the trials will be suppressed during the averaging process. In practical terms, this means that the stronger the signal that should be measured, the less trials are necessary to achieve a reasonable reduction in the signal-to-noise ratio. Luck [69] suggests a minimum of 30–60 trials if the experiments deals with larger components such as the P3, and 150–200 trials if smaller components such as the N2 are involved. These figures correspond to those provided by [46]; 150 trials using smaller components and 36 trials for stronger components.

The *interstimulus interval* (*ISI*) is defined by the time between two trials and can either be a fixed time (e.g., 1,200 ms) or can be jittered randomly within a certain time frame (e.g., between 1,000 and 1,400 ms). Following the recommendation of [46] a short fixed ISI of 500 ms can be used for small and early components (e.g., MMN). For later and stronger cognitive components (e.g., P3), a fixed ISI of 1–2 s is recommended. For cognitive components, it should be considered that the ISI needs to be longer when the experimental task is difficult [46].

The term "epoch" is used for a time interval surrounding an experimental event. If an ERP caused by the onset of a certain sound should be analyzed, an epoch can, e.g., be defined by the time interval −200 to 1,000 ms, whereas 0 ms denotes the actual onset of the stimulus.

1.3.4.3 The Component Concept

The analysis of the event-related potentials can be done without prior knowledge by comparing ERPs recorded under different task conditions. In accordance with [71], this can be done by extracting information about the *timing* (time course), the degree of engagement (*amplitude*), and functional equivalence of the underlying cognitive process (distribution across the scalp). When ERPs were recorded under two experimental conditions, it was possible to check the difference between the aforementioned parameters (see Fig. 1.3 for the procedure how to quantify ERP components).

In the event that prior knowledge about the intended brain responses has been given, additional information about where the signal was generated and what cognitive processes were associated with it can be gained. This is done by analyzing different components which constitute the ERP waveform (such as the P300). What the standard definition of the term "component" should be, is still under discussion [69]. As it is still extremely difficult to isolate components, several considerations concerning the experimental set-up and analyzes should be taken into account (see [69] for a detailed explanation on how to design ERP experiments).

Firstly, it is not always certain that measurement peaks correspond to the intended components, and secondly, it cannot be presumed that the average ERP waveform is

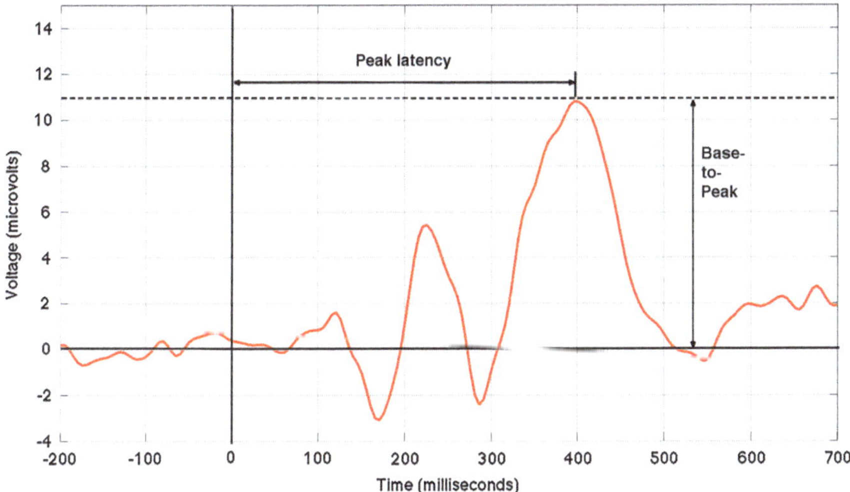

Fig. 1.3 Procedures for quantifying ERP components. Representation of an ERP waveform (based on [49]). Illustrated are two procedures: **1** base-to-peak, which corresponds to the voltage at the peak of the component and **2** peak latency, which corresponds to the time at the peak

similar to the single-trial waveform. Strategies to prevent misinterpretations should focus on specific, well-studied, large, and isolated components.

For the analysis of ERPs, a small set of electrodes can be sufficient, usually up to 8 electrodes; they should be distributed along the central line following the *10/20* system [65], for hemispheric differences, equally distributed electrodes over the right and left hemispheres are advisable. More electrodes are needed for the analysis of more complex patterns, e.g., spatial pattern distribution.

As evoked potentials depend on exact timing, it is important that triggers are exactly synchronized to be able to average the signal while keeping the temporal information intact. ERPs cannot be observed in the raw EEG, as they are overshadowed by other unrelated activity which disappears after averaging several trials of single ERP recordings.

Usually 20–30 trials as a minimum are needed for an average ERP per stimulus class; baseline averaging is carried out using the average value of the voltage in the interval, usually up to 200 ms prior to the stimulus. This rather high number of trials compared to standard quality tests also explains the usually small number of participants used for EEG studies.

The aforementioned averaging methods are performed offline and as an average across a group of participants. This average between all participants is the grand average and is the result most often plotted in these studies. Using classification techniques, this can be transferred to the online analysis of incoming physiological signals, meaning the decision as to whether the occurring brain activity of the preceding stimulus has been evoked by one special class of stimuli [72]. In the case of Quality of Experience, an exemplary class of degradation can serve this purpose.

With classification as an indicator of separability, distinctions can be drawn between perceived stimulus classes. For a tutorial on single-trial ERP classifications see [73]; a more detailed description of ERP classification will be given in Sect. 1.3.4.5.

1.3.4.4 Sensory Components

Mismatch Negativity (MMN): *Mismatch Negativity* (*MMN*) is a measure of low-level visual and auditory memory [74]. It is an automatic process caused by differences between the currently processed stimulus and previously received stimuli which generated an internal sensory reference [75]. It is elicited during the range of 100–250 ms after stimulus onset [46]. This automatic process is not conscious and can also be found in sleeping participants [76]. Näätänen et al. (2007) first described the MMN and explained it as follows: "The 'traditional' MMN is generated by the brains automatic response to any change in auditory stimulation exceeding a certain limit roughly corresponding to the behavioral discrimination threshold" [77]. This could indicate that the MMN—as an initial automatic response—could be the most sensitive for measuring physiological responses evoked by stimuli of various speech quality. Nevertheless, due to the smaller signal amplitude it is more difficult to gather reliable results and more trials are needed to increase the signal-to-noise ratio. The review from Garrido et al. (2009) gives a recent overview of the MMN [75].

Especially the N100 and P200 appear to be very meaningful for audiovisual integration; Pilling (2009) [78] reasoned that the N1/P2 amplitude reduction due to audiovisual synchrony represents a marker of audiovisual integration. These results are very interesting for QoE research concerned with analyzing the perception and quality of audiovisual stimuli. This is due to the fact that—during the transmission of audiovisual material—the alignment of speech and visual information can be one important source of degradation. An ERP component which represents a marker of synchrony could be used as an additional source of measurement for quality testing. As this book investigates speech exclusively, the investigation of this phenomena must be reserved for future research.

Folstein et al. (2008) compare the advantage of cognitive control and mismatch in the N2 component [79]. The authors argue that the N2 component is not necessarily evoked by mismatch exclusively and can also comprise regulation of strategy and feedback, which can in turn be used for the regulation of the strategy used for an experimental task.

The N200: The *N200* can be observed for the auditory modality (frontal or central recording sites) and reflects a mismatch between stimulus features, or between stimulus features and a previously formed template [49]. In contrast to the MMN, the N200 is evoked if the attention of participants is directed towards the stimulus. Therefore, templates can be formed by the participants themselves, as instructed during an experiment.

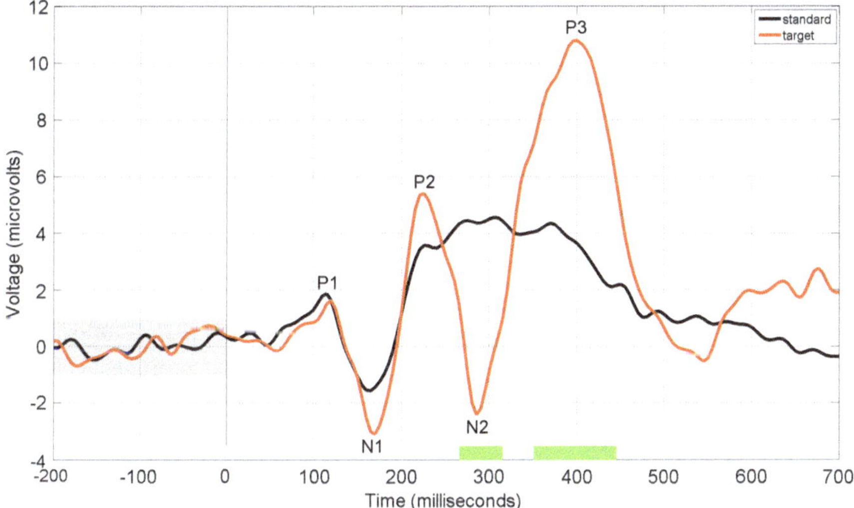

Fig. 1.4 Two *Event-Related Potentials* (*ERPs*) evoked by auditory stimuli (grand average, electrode Cz). *Oddball paradigm* with two tones as stimuli was utilized; *standard* (440 Hz beep tone for 150 ms, probability of 80 %, 480 repetitions) and *target* (1,000 Hz beep tone for 150 ms, probability of 20 %, 120 repetitions); the participants had to click a button as quickly as possible to indicate which tone they heard last. *Interstimulus interval* was set at 1,500 ms and sampling rate at 200 Hz. Data was band-pass filtered (0.1–40 Hz) and only correctly identified trials were used for display. The *gray bar* indicates the time interval used for baseline correction (−200 to 0 ms: where 0 ms is the stimulus onset) and the *green* interval indicates the intervals where the two ERPs are significantly different, running t-test with adjusted $p < 0.05$. Components are marked as P1-3 for occurring positive peaks and N1-2 for negative peaks. Reprinted with kind permission from Springer Science+Business Media: Antons et al. [5]

1.3.4.5 The Late Cognitive ERPs

In contrast to the early sensory ERPs, which are mainly related to sensory processing, late cognitive potentials are mainly evoked by higher-level cognitive processes, e.g., a detection task such as identifying a certain speech stimulus.

The P300: The *P300* component, also referred to as P3, is a positive peak approximately 300 ms after stimulus onset. Examples of spatial distribution on the scalp and for the time course can be found in Figs. 2.5 and 1.4, respectively.

The component is divided into two parts: P3a and P3b. P3a is the result of a comparison between newly perceived information and internal memory copies, similar to the MMN. The P3b component is elicited by task-related attention. In general, the P300 is elicited when a deviant stimulus is presented among a series of more frequent "regular" stimuli, e.g., a high tone among a repeated series of low tones, which is one of the standard tests in ERP research called the "oddball paradigm". The review from Polich (2007) provides background information on all relevant processes behind the P300, P3a, and P3b components [80]. Even later components such as the N400 are associated with the semantic processing of stimuli, e.g., on a sentence level.

Guidelines from Duncan et al. (2009) supply practical advice for procedures to measure MMN, P300, and N400 [46].

Using this methodological approach, recent neurophysiological studies of auditory processing have led to models for auditory processing and the conscious perception of stimulus features [81]. Koelsch (2009), who investigated early components of music processing, provides an overview on the connection of music processing and MMN, which are both early stimulus processing stages, though triggered differently [82]. Furthermore, a first experiment using degradation classes, which are of interest for research in the telecommunication industry, was conducted by Miettinen et al. (2010) in context of magnetoencephalography (MEG), where researchers were able to demonstrate a significant increase in the measured amplitudes of distorted stimuli [24]. In preliminary publications, it could be shown that a certain electroencephalographic (EEG) technique, event-related potential (ERP) analysis, is a useful and invaluable tool in quality research [1, 41, 42, 43]. In these studies, it was shown that (1) the easier the task to detect a degraded speech signal was, the earlier and higher the P300 peak occurred, and (2) that the classifier cannot merely discriminate between behavioral hits and misses, but also finds P300-like patterns within the trials marked as misses. A proportion of stimuli that were missed on the behavioral level were presumably still processed non-consciously.

The technique of measuring event-related potentials, especially P300 peak parameters such as amplitude and latency, will be applied in Chaps. 2, 3 and 4.

The N400: The *N400* is mainly evoked by linguistic processes [46] and [49]. In the experimental task, sentences have to be read by the participants (each single word in a sequence). While most of the sentences end with a semantically correct ending, some are presented with a semantically incorrect ending. The latter ones evoke a negativity response around 400 ms after stimulus onset—the appearance of the last word of the sentence—called N400. The amplitude of the component is dependent on the strength of semantic mismatch [49]. In addition to the above mentioned reading task, N400 components can be evoked by spoken and signed language as well [46].

Classification of Single-Trial ERPs: If not only the average neuronal response of one participant or a group of participants is of interest, EEG data can be classified. Usually, the types of classifiers can be either *unsupervised* or *supervised*; in supervised cases, class labels (e.g., high and low quality) are known during the training phase of the classification parameters. As only *supervised* classifications have been used in the analyses of this book, the following sections only discuss supervised classification.

Classification in the domain of BCI research is usually undertaken by assigning a single-trial allocation to one of two given classes [83]. In the case of an experiment where speech signals with two quality levels (high and low quality) were used, this would result in a prediction if the EEG data used during the trial was evoked by a high-quality stimulus or the stimuli with low quality. The advantage of the ERP technique, i.e., that non-event-related signal components are suppressed (see Sect. 1.3.4.2) in the averaging process, is not true for single-trial analysis.

As input (*feature*) for the classifier, similar parameters as used for the analysis of P300 components, namely, *peak amplitudes and latencies*.

Classifying EEG data has the advantage that new incoming physiological data can be assigned to a class after a training phase, without asking the participants. In an EEG test, this has the advantage that the presentation of stimuli would not have to be interrupted in order to ask participants for their opinion. In addition to this—due to the high sensitivity of classifiers—even small variations can be detected, and therefore, minimal changes of neuronal pattern can be associated with stimulus properties such as quality.

Many different classifiers can be used for performing the classification task such as *Support Vector Machines* (*SVM*) or *Linear Discriminant Analysis* (*LDA*)—however, the LDA method has been proven to be one most successful techniques for the classification of ERP data [73]. The classification performance is frequently measured in terms of balanced accuracy, expressed as the *Area Under the Curve* (*AUC*) of the *Receiver Operating Characteristic* (*ROC*), see Eq. 2.1 in Chap. 2 and [84]. Balanced accuracy stands for the relationship defined by true positive (tp) rate and false positive (fp) rate of a 2-class problem, including the true negative (tn) rate and the false negative (fn) rate. A value of AUCb > 0.9 reflects excellent classification and AUCb = 0.5 chance level.

In this book, LDA will be used for the classification of ERP patterns in Chaps. 2 and 3.

1.3.4.6 Findings Related to QoE

A first experiment using degradation classes which are of interest to research in the telecommunication industry, was conducted by Miettinen et al. (2010) using *magnetoencephalography* (*MEG*), where scientists were able to demonstrate a significant increase in the measured amplitudes of distorted stimuli [24].

This measurement technique was further developed for (audio-)visual stimuli. In the studies conducted by Arndt et al. (2012), the *two-alternative forced choice* (*2AFC*) approach was used, which is a reduced implementation of the *double stimulus continuous quality scale* (*DSCQS*) (see [17]). Here, an initial video sequence with the reference stimulus was immediately followed by a possibly distorted one. As a distorted element in these experiments, artificial blockiness was introduced and varied in block length. Participants had to tell after each trial whether they had perceived a distortion in the second part or not. Findings from previous auditory studies could be repeated, and the same relation for visual stimuli was demonstrated: the P300 is more distinct with more distorted stimuli [44]. In a next step, bi-modal stimuli were introduced using the *2AFC* paradigm [85]. Besides the already established relationship between P300 amplitudes and distortion levels, in this experiment a significantly higher relation between *Mean Opinion Score* (*MOS*) [19] and obtained P300 amplitudes was observed. In other studies with an exclusively visual basis, Scholler et al. confirmed these findings and were also able to show that the ERP of stimuli perceived as not degraded at the subjective level could be identified similarly to those perceived as degraded [45]. Another experiment using visual stimuli was conducted by Lindemann et al. [86]. Here, rather different kinds of distortions than intensities were examined. Researchers were also able to demonstrate high classification rates between distorted and undistorted stimuli with the obtained data.

New technologies such as 3D videos can also be examined in respect to their qualitative and visual discomfort. This was undertaken in a experiment by Li et al. (2008), in which scientists were able to observe higher visual discomfort (1) while watching 3D content versus 2D content, and (2) while watching longer 3D sequences versus shorter ones [87].

1.4 Outline and Objective of this Work

This book summarizes the work of the author on neural correlates of quality perception for complex speech signals. Some of the results of this book have been previously published [1, 2, 3, 4, 41] and/or presented as contributions to Study Group 12 of the International Telecommunication Union [88, 89].

Besides previously published aspects of its results, this work presents a complete picture, overview, and restructured discourse concerning how neuronal signals can be used as complementary measures of speech quality perception assessment.

The structure of this book has been determined by the length of the stimulus material discussed here from phonemes to audiobooks. Figure 1.5 depicts the main structure of the book and will guide the reader through the experiments which have been conducted.

In respect to stimulus lengths, phonemes, words, and sentences, one degradation class was selected (signal-correlated noise, bit rate reduction of a speech codec, and reverberation).

As can be seen in the selection of physiological techniques and degradation classes in this book, one experiment with one degradation class, one stimulus length, and one physiological technique was carried out. As the main focus is the development of a combined test set-up for subjective and EEG methods, a set-up was used with selected stimuli and suitable methods. The intent was to show that the set-up used here functions with different stimuli and rather than testing all possible combinations, which remains a task for further research.

Chapter 2 deals with the implementation of an ERP technique for speech quality assessment, generally with the use of short stimuli, i.e., phonemes in combination with a generic degradation (signal-correlated noise with varying signal-to-noise ratios); from this point onwards, this experiment will be referred to as "the Phoneme Experiment". The second Chapter will also deal with the question about whether the length of a presented speech signal has an influence on the resulting subjective quality rating. In Chaps. 3 and 4, the technique developed here will be applied to stimuli with the length of words and sentences, respectively. In these studies, the impaired quality was due to the reduced bit rate of a speech codec and reverberation; these experiments will be referred to as "the Word Experiment" and "the Sentence Experiment", respectively. Chapters 5 and 6 deal with the implementation

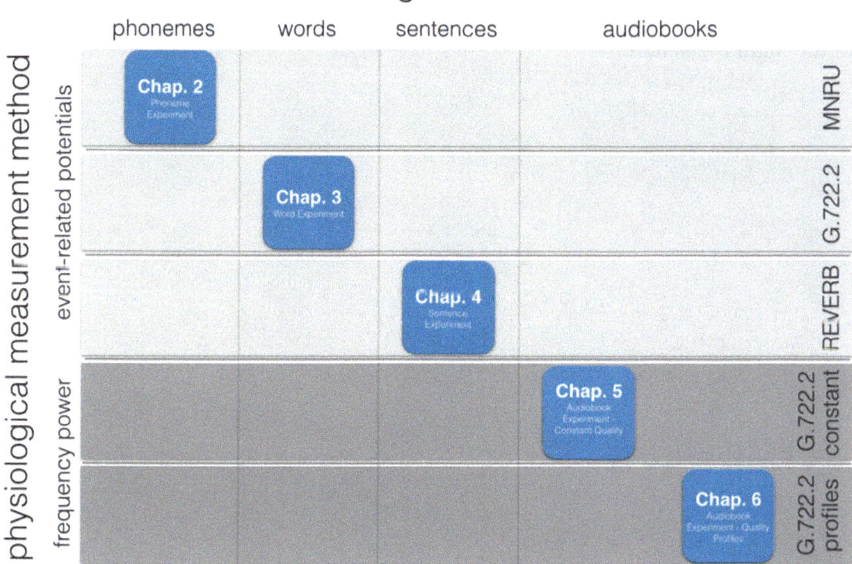

Fig. 1.5 Conducted experiments and structure of this book. Different lengths of stimuli on the x-axis (phonemes, words, sentences, and audiobooks). Physiological measurement techniques on the y-axis (EEG frequency band power and event-related potentials). Applied classes of degradations are color-coded (*grey bars*) and indicated on the *right* (signal-correlated noise introduced by a modulated noise regulation unit (MNRU), bit rate reductions introduced by using different settings of a speech codec in accordance with ITU-T Rec. G.722.2, and reverberation (REVERB) introduced by different room impulse responses

of the band frequency power method, showing the impact of quality variations on the cognitive state of listeners under constant and varying quality conditions; these chapters will be referred to as “the Audiobook Experiment—Constant Quality” (Chap. 5) and “the Audiobook Experiment—Quality Profiles” (Chap. 6), respectively.

In Chap. 7, the results of all studies will be discussed and topics for future research will be identified.

The main intent of this work is to utilize physiological techniques which can be added as complementary measurements during standard subjective speech quality experiments. Thus, although physiological measuring techniques are still under development for speech quality research, promising advances for real-life and standardized applications can be identified.

As a general remark on the application of extracted physiological parameters, it must be mentioned that the objective of this book is to present the application of existing techniques—namely analyzing the ERP components and frequency band power of EEG bands—within a field related to speech quality research. For this reason, the analysis of the data presented here will predominantly focus on how the derived parameters can be utilized as an extension of a standard quality test. New findings in physiological methodology and the underlying analytical foundation are not the main focus here.

Chapter 2
ERPs and Quality Ratings Evoked by Phoneme Stimuli Under Varying SNR Conditions

In this chapter, the first and second contributions (see Sect. 1.4) of this book will be introduced. First, a test set-up combining neurophysiological and subjective quality assessment methods for speech quality perception testing will be presented. Secondly, the functionality of this set-up will be validated for short speech stimuli with the length of phonemes and a generic quality impairment, i.e., signal-correlated noise (for an overview see Fig. 2.1). It will be shown that a test set-up combining neurophysiological and subjective quality assessment methods for speech quality perception testing is suitable for measuring the perceived speech quality, and in some instances (trials) it is advanced in comparison to a standard subjective test set-up.

In this first experiment, the selection of short auditory stimuli is based on the fact that—in the majority of neurophysiological speech/audio research—stimuli are of short duration (up to approx. 2 s). In order to be able to follow standard neurophysiological recommendations (see Sect. 1.3.4) for ERP experiments, the duration of stimuli was restricted to the length of phonemes. The degradation class was selected on the basis that, in the first validation of the test set-up, a continuous degradation should be used. Signal-correlated noise, as used for comparisons of different transmission systems [90], was existent during the entire speech signal, but only when the speech signal was active (no noise in the pauses).

The test set-up and its initial validation will be described below. In Sect. 2.2, the experimental set-up including EEG, ERP measurement, and opinion tests will be explained, followed by a global analysis (Sect. 2.3), an introduction to the statistical tools (Sect. 2.4), and a presentation of the experimental results (Sect. 2.5). In the following chapters the test set-up will be extended to include stimuli with longer duration and other quality reductions (see Chap. 3 for stimuli with word length and bit rate reduction; Chap. 4 for stimuli with sentence length and reverberation).

J.-N. Antons, *Neural Correlates of Quality Perception for Complex Speech Signals*,
T-Labs Series in Telecommunication Services, DOI 10.1007/978-3-319-15521-0_2

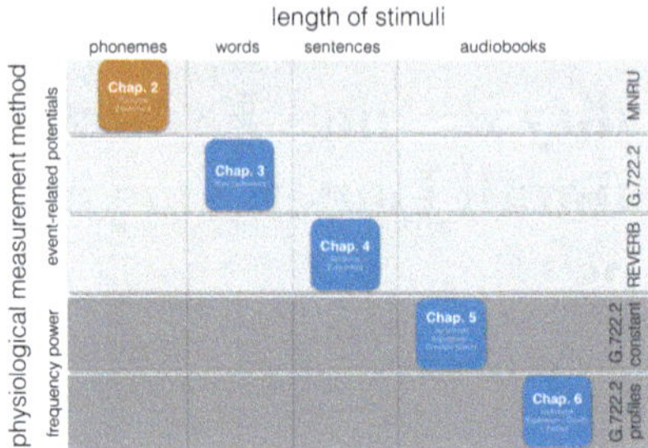

Fig. 2.1 Conducted experiments and structure of this book. Different lengths of stimuli on the x-axis (phonemes, words, sentences, and audiobooks). Physiological measurement techniques on the y-axis (EEG frequency band power and event-related potentials). Applied classes of degradations are color-coded (*grey bars*) and indicated on the *right* (signal-correlated noise introduced by a modulated noise regulation unit (MNRU), bit rate reductions introduced by using different settings of a speech codec in accordance with ITU-T Rec. G.722.2, and reverberation (REVERB) introduced by different room impulse responses. Current chapter is indicated in *orange*. Chapter 2, *Phonemes Experiment*

2.1 Introduction

In the *Phoneme Experiment*, *standard* and *deviant* stimuli are presented in terms of the *oddball paradigm* (see Sects. 1.3.4.4 and 2.2.3), in which the phoneme /a/, is uttered by a male speaker and presented continuously in a high-quality (HQ) version, interrupted by a disturbed version of that phoneme.[1] As distortion, signal-correlated noise generated by a Modulated Noise Reference Unit (MNRU) [19], was chosen. This degradation is well suited for an initial check of the test set-up, as noise is one of the most ubiquitous factors that hamper efficient communication [39] and will therefore most definitely have an impact on quality perception. In addition to its impact on quality perception, the MNRU is recommended by the International Telecommunication Union (ITU) and can be used to compare, e.g., different speech transmission systems. As a well known degradation factor with an almost guaranteed influence on quality perception, signal-correlated noise was selected to achieve the degradation effect. The extent of the distortion was varied at four levels, i.e., from LQ1 to LQ4, where LQ1 (low quality 1, LQ1) referred to the weakest distortion. It was hypothesized that the P300 peak amplitude and latency (see Sect. 1.3.4.3) would vary as a consequence of distortion intensity. In addition to the distorted /a/, a second deviant (/i/) was presented as a control stimulus or "sanity check". This stimulus should cause a P300 under all circumstances.

[1] This chapter is based on a previous publication; text fragments, tables, and figures are based on Antons et al. [1]. Reprinted, with permission, from [1].

2.2 Methods

2.2.1 Participants

Ten right-handed students and personnel from the Technical University of Berlin participated in the experiment (six females, four males; average age = 28.20 years; SD = 8.49; range = 19–51 years old), all of them native German speakers. All participants reported normal auditory acuity and no medical problems. Handedness was assessed using an inventory from Oldfield (1971) [91]. Participants gave their informed consent and received monetary compensation. The experiments were conducted in accordance with ethical principles that have their origin in the Declaration of Helsinki.

2.2.2 Material

Fourteen vowel phonemes were used: /a/ undisturbed, /i/ undisturbed, and twelve disturbed versions of /a/ impaired with signal-correlated noise. None of these phonemes have lexical meaning in German. The vowel /a/ was selected due to the fact that it has an clear and strong onset, and in addition to this, a high energy expenditure. The vowel /i/ was selected, as it should be clearly distinguishable from the preceding vowel /a/. In order to account for possible individual differences in hearing sensitivity, a set of stimuli was selected for each participant individually, based on her/his detection rate. Out of the stimulus set as a whole, an individual sub-set of four stimuli was selected for each participant, based on the results of a pre-test. During the pre-test all fourteen stimuli were presented to the participants four times in the context of an opinion test. The task of the participants was to rate a stimulus as belonging to one of two classes, high quality (no degradation) and low quality (with degradation). A detection rate was calculated by dividing the number of correctly identified low-quality stimuli by the total number of stimuli presented in the corresponding category. Based on the resulting detection rate, a final stimulus selection of four stimuli was carried out. For every participant, those stimuli were selected that were closest to the targeted detection rates of 100, 75, 25, and 0 % for the four selected stimulus levels. The correlated *signal-to-noise ratios (SNR)* for the complete stimulus set were set at: 14, 16, 18, 20, 21, 22, 23, 24, 25, 26, 28, and 30 dB. Stimulus material was digitally recorded in a sound-attenuated experimental chamber with a 48 kHz sampling rate. The phonemes were articulated numerous times by a male speaker. In order to keep the acoustic variability minimal, only one version of each phoneme was selected. Intensities were normalized using the root mean square of the speech period in the sound file using the software Adobe Audition®. The duration of each stimulus was set at 200 ms. The stimuli were degraded by a MNRU according to ITU-T Rec. P.810 in a controlled and scalable way [90]. The median SNR for the deviant stimuli and for all participants can be found in Table 2.1.

Table 2.1 SNR(db) for all participants and median SNRs

Participant	SNR(dB)				
	HQ	LQ1	LQ2	LQ3	LQ4
1	100	28	24	21	5
2	100	25	20	17	5
3	100	28	24	22	5
4	100	24	22	21	5
5	100	30	26	22	5
6	100	35	28	26	5
7	100	30	26	22	5
8	100	22	20	18	5
9	100	28	25	22	5
10	100	30	25	20	5
Median	100	28	24	21	5

2.2.3 Experimental Design and Procedure

Under these experimental conditions, *oddball stimulus sequences* equalling 300 trials in total were presented (see Fig. 1.2 for a visualization of the oddball sequence). In each sequence, the undisturbed phoneme /a/ served as the *standard* stimulus (70 % of the trials), whereas the undisturbed phoneme /i/ as well as four selected disturbed versions of the phoneme /a/ served as *deviants* (6 % of the trials, respectively), delivered in a pseudo-randomized order, forcing at least one standard to be presented between successive deviants. As the oddball paradigm has not been frequently used for research studies in the telecommunications field concerned with quality so far, a control stimulus (/i/) was initially included as a sanity check, using a well-established P300 event [80]. An exploration of the P300 evoked by the /i/ stimulus showed indeed that a novelty P300 was consistently evoked in all participants. As a P300 for at least one degradation condition was eventually found for every participant, a further evaluation of the control stimulus (i.e. /i/) was not conducted. Per participant, eight to twelve sequences were recorded. All of these sequences contained six trials per degradation strength (SNR), respectively. Based on the behavioral results of each participant during the pre-test, an individual set of four stimuli was chosen for the experiment. As already mentioned, it was hypothesized that the four selected degradation levels that were selected would be detected with a rate of LQ4 = 100 %, LQ3 = 75 %, LQ2 = 25 %, and LQ1= 0 %. Stimuli were presented with an interstimulus interval varying from 1,000 to 1,500 ms (time between two consecutive stimuli). Participants were seated comfortably and instructed to press a button whenever they detected one of the deviants or the control stimulus (identification task, LQ1-4 and /i/). Stimuli were presented binaurally at the listening level preferred by the individual through Sennheiser® in-ear headphones. After the pre-test and physiological measurement, participants additionally had to rate all 12 stimuli—the complete stimulus set—in

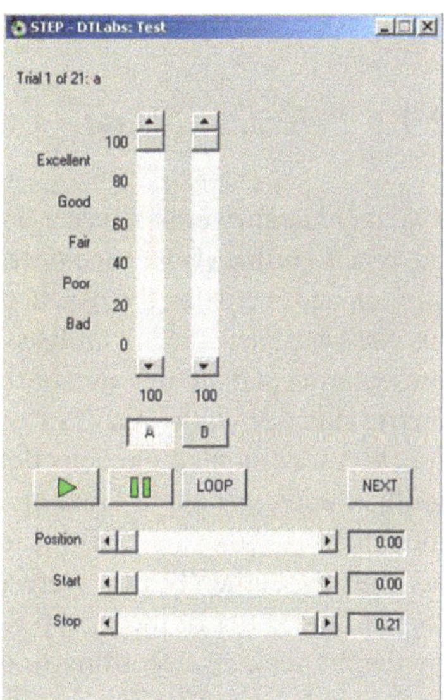

Fig. 2.2 Graphical user interface of the STEP software. Used to collect the subjective quality ratings of participants on a continuous quality scale (CQS), ranging from bad (0) to excellent (100). An adapted version—without hidden references and anchors—of the MUlti Stimulus test with hidden reference and anchor (MUSHRA) methods

respect to their quality. For this test, the *MUlti Stimulus test with Hidden Reference and Anchor* (*MUSHRA*) techniques in accordance with ITU-R Rec. BS.1534-2 [33] was adapted to match as closely as possible the EEG experiment that was carried out. Therefore, no hidden reference and no anchor stimulus were used. Two stimuli were visually presented at the same time and had to be rated on a *continuous quality scale* (*CQS*) ranging from bad (0) to excellent (100). The Audio Research Lab software STEP (see Fig. 2.2) was used for collecting the data.

An experimental session lasted approximately 3 h (plus additional time for electrode application and removal), including breaks to avoid participant fatigue.

2.2.4 Electrophysiological Recordings

The EEG (Ag/AgCl electrodes, Brain Products GmbH, Garching, Germany) was recorded continuously using 64 standard scalp locations according to the extended 10–20 system (AF3-4, 7-8; FAF1-2; Fz, 3-10; Fp1-2; FFC1-2, 5 8; FT7-10; FCz, 1-6; CFC5-8; Cz, 3-6; CCP7-8; CP1-2, 5-6; T7-8; TP7-10; P3-4, Pz, 7-8; POz; O1-2, and the right mastoid) [65]. The reference electrode was placed on the tip of the nose. Electroocular activity was recorded with two bipolar electrode pairs. Impedances were kept below 10 kOhm. The signal was digitized with a 16-bit resolution and a sampling rate of 1,000 Hz.

2.3 Data Analysis

2.3.1 Behavioral Data

Two parameters were derived as behavioral data during the EEG measurement. First, the reaction time (from tone onset to button press in milliseconds) for the different stimuli, and secondly, the psychometric functions. The reaction time for each stimulus class was measured in milliseconds, as the duration between the onset of stimulus presentation and the reaction of the participant (received button click). The psychometric function is the result of the detection rate as a function of SNR. A logistic function was fitted to the detection rates of the stimulus levels with the MATLAB® toolbox *psignifit*, approximating the data points in accordance with least-squares models [40]. After the EEG measurement, participants had to complete an opinion test, more specifically, they were asked to rate all LQ levels on a scale from excellent to bad. The slider of the Audio Research Lab software STEP (see Fig. 2.2) was set by the participants according to the *continuous quality scale (CQS)* in compliance with ITU-R Rec. BS.1534-2 [33] from excellent (100) to bad (0).

2.3.2 ERP Data

Off-line signal processing was carried out using the MATLAB® toolbox EEGLAB [92]. The raw EEG data were low-pass filtered with a finite impulse response filter (low-pass filter with a critical frequency of 40 Hz). EEG epochs—the time interval around one stimulus, as well as one trial—with a length of 1,400 ms, time-locked to the onset of the stimuli, and including a 200 ms pre-stimulus baseline, were extracted and averaged separately for each condition (HQ, LQ1-4 and C) and for each participant. Epochs (−200–1,200 ms around stimulus onset) with an amplitude change exceeding 100 microvolt at any of the recording channels were rejected as artifacts, as this voltage change is unlikely to be produced by neuronal activity. Grand averages were subsequently computed from the individual participant averages. To quantify the deviance-related effects of P300, the peak latency and peak amplitude were measured in a fixed time frame relative to the pre-stimulus baseline (see Fig. 1.3). The time frame for P300 quantification was set from 200 to 1,000 ms after stimulus onset. The maximal positive amplitude in this time frame was automatically determined; its voltage and latency were extracted for further analysis.

2.3.3 Classification

The aim of classification was to identify trials in which the participant was not able to detect a degraded stimulus, although an activation pattern similar to conscious detection was present. The detailed selection of classes can be found in Sect. 2.4.3 below. The classification was done using the MATLAB® toolbox BCILAB [93]. The comparison of ERP data with classifications is usually done by comparing the HQ versus the LQ ERPs. Features were the averaged voltages for the time windows 200–400, 400–600, 600–800, and 800–1,000 ms, for all EEG channels. In case of equal covariance matrices for both classes and Gaussian distributions, *Linear Discriminant Analysis* (*LDA*) is the optimal classifier [94]. In respect to ERP signals, LDA is most suitable for classification purposes (for detailed information on single-trial classification of EEG data see [73]). An LDA with automatic regularization of the estimated covariance matrix and utilizing shrinkage procedures was applied.

2.4 Statistical Analysis

2.4.1 Behavioral Data

A Milton Friedman Test with a post-hoc comparison was calculated for reaction times [95]. As regards the opinion test, an analysis of variance (ANOVA)—with *degradation intensity* as the independent variable and the *mean opinion score* (*MOS*) as the dependent variable—was also calculated [96].

2.4.2 ERP Data

Deviance-related effects, namely, the presence and amplitude of P300 responses, were analyzed on the basis of data from the Cz electrode where P300 is typically at its maximum. Whereas in the present experiment 64 electrodes were used, the long-term goal for future research is to identify a minimal electrode placement providing a reliable response estimate in the majority of participants. Accordingly, in a pre-analysis a grand average was calculated and the one single electrode exhibiting the mean maximal P300 amplitude was identified. As this was found at the vertex,[2] all further analyses were carried out using the Cz electrode. Figure 2.4 shows exemplary ERPs for different stimuli classes. In order to test the presence of P300 under controlled conditions (/i/), deviant responses were compared to the corresponding standard responses, evoked by the undisturbed standard phoneme, by means of dependent t-tests. The minimum number of epochs constituting the ERP was set at 25. The peak latency and peak amplitude of the P300 responses were analyzed by means of repeated measures ANOVA with the factor *stimulus* (*HQ, LQ1-4, and C*). Finally, post-hoc comparisons were drawn between target types pairwise with Sidak-adjusted alpha levels.

[2] The uppermost surface of the head.

2.4.3 Classification

Classification was carried out participant-wise using bandpass-filtered raw data (0.2–7 Hz). Each LQ class was further divided into two separate subclasses: hits (true positives) and misses (false negatives). Stimuli that were degraded and not detected by the participants were labeled as misses. Detected degradations were labeled as hits. Two classifications were completed: (1) the training of a classifier to distinguish between hits and correctly reported HQ trials and testing this classifier on the same events (HQ against hits of each class), (2) once again the training of a classifier to distinguish between hits and correctly reported HQ trials, but alternatively, later testing on misses versus correctly reported HQ trials. For training purposes in the second classification, one half of the HQ trials as well as the hit trials of each LQ class were utilized. Two separate sets of HQ trials (HQ1 and HQ2) were created for the second classification by selecting even and odd HQ trials and assigning them to HQ1 and HQ2, respectively. For testing purposes, the other half of the HQ trials and the missed trials were used. This approach was first introduced in [42]. The analysis was carried out for all stimulus levels with a 5-fold cross-validation. Only if a minimum of 15 trials containing hits (classification 1) or containing 15 hits and 15 misses (classification 2) was available, would classification be carried out (minimal number of trials needed to train and test a classifier). The classification of hits within each target class versus HQ demonstrates that the classification of neural reactions based on the perception of degraded stimuli is feasible. The second analysis, classifying misses versus HQ trials, revealed differences in the EEG signals due to degradations which were not noticed at the behavioral level. Nonetheless, these two classes probably differ on a physiological level, as the degradation is still processed at the neural level. Classification performance was measured in terms of balanced accuracy (see also Sect. 1.3.4.5), expressed as the *Area Under the Curve* (*AUC*) of the *Receiver Operating Characteristic* (*ROC*), see Eq. 2.1 [84].

$$AUCb = \frac{\left(\frac{tp}{(tp+fn)} + \frac{tn}{(fp+tn)}\right)}{2} \tag{2.1}$$

Balanced accuracy stands for the relationship defined by true positive (tp) rate and false positive (fp) rate of a 2-class problem, including the true negative (tn) rate and the false negative (fn) rate. A value of AUCb > 0.9 reflects excellent classification and AUCb $= 0.5$ chance level. The significance level of the classification outcome was tested—as to whether the classification result was significantly different to chance occurrence—using a Wilcoxon rank-sum test.

2.5 Results

2.5.1 Behavioral Data

For the opinion test, the *ANOVA* with *Degradation Intensity* as the independent variable, and the *Mean Opinion Score* (*MOS*) as the dependent variable in the opinion test data, proved to be a significant influence on the factor *Stimulus* (strength of degradation) ($F(131{,}4404) = 306.64, p < 0.01, \eta^2 = 0.76$). The post-hoc test (Sidak adjustment for pairwise comparisons) attained significance at a level of 21 dB ($p < 0.05$) compared to non-degraded stimuli. The psychometric function fits for all participants have been plotted in Fig. 2.3.

The mean reaction times for the different conditions can be found in Table 2.2.

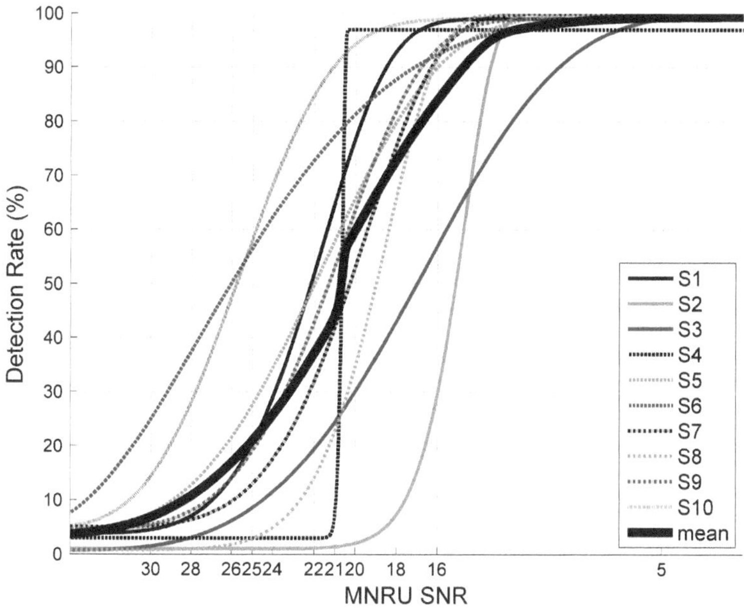

Fig. 2.3 Psychometric function fits from the psychophysical data; for all participants (participant (S) 1–10) and the average across all participants. Reprinted, with permission, from [1]

Table 2.2 Mean reaction times in milliseconds for all levels of degradation

	Milliseconds			
	LQ1	LQ2	LQ3	LQ4
Mean	724.93	736.28	749.24	598.54

LQ1 represents the weakest degradation and LQ4 the strongest

The reaction time for LQ1-3 are on a similar level, but significantly different compared to LQ4 ($p < 0.05$). For stimulus condition LQ4, the reaction time was shorter.

2.5.2 ERP Data

The time frame for *P300 peak* quantification was set at 200–1,000 ms after stimulus onset; within this interval, the maximum value of the ERP curve was identified and *peak latency and amplitude* were extracted as parameters. In order to test for the general presence of a P300 response evoked by the test set-up, it was checked whether the control stimulus (/i/) evoked a significantly different peak amplitude in comparison to the P300 peak amplitude of the standard (HQ, /a/ in high quality). As these two stimuli are clearly distinct on a physical level, and easily distinguishable if participants were asked to tell which stimulus had been presented, a clear neuronal response was anticipated. As regards testing to determine whether the difference defined by the *P300 peak amplitude* of the grand average was in fact significant, a two-tailed dependent t-test was performed. The t-test result was clearly significant ($t = 6.37$, $p < 0.01$). Due to the fact that the usage of the stimulus /i/ was only intended to check whether the test set-up was suitable for evoking a P300 response, the results connected with this stimulus will not be further pursued.

The ANOVA for repeated measurements revealed a significant main effect on the factor *Stimulus* ($F(9{,}27) = 3.56$, $p < 0.01$, $\eta^2 = 0.54$). Figure 2.4 shows the grand average ERPs, the arrows indicate the location of the *P300 peak* for each LQ. For dependent variable *P300 peak amplitudes*, a significant effect was found ($F(3{,}9) = 11.34$, $p < 0.05$, $\eta^2 = 0.79$), as well as for the dependent variable *P300 peak latency* ($F(3{,}9) = 9.35$, $p < 0.05$, $\eta^2 = 0.75$). The pairwise comparison (Sidak adjustment for pairwise comparisons) revealed a significant difference between LQ2 and LQ4 for the *peak amplitude* ($p < 0.05$). A significant effect could be found between LQ2 and LQ3 for the *peak latency* ($p < 0.05$), in addition to a significant effect between LQ2 and LQ4 for the *peak latency* ($p < 0.05$). Figure 2.5 shows the scalp distribution of voltage for the different stimulus conditions (hits and correct rejections for LQ1-4 and HQ, respectively).

For LQ4, a broad reaction was detected. For the less disturbed stimuli, a reaction was provoked, but not as strong as for LQ4. In addition, a correlation between the *P300 amplitude* for electrode Cz and the detection rate was found ($r = 0.42$, $p < 0.05$). Within the ERP data, a negative correlation between the *P300 amplitude* and the *P300 latency* at electrode Pz could be observed ($r = -0.33$, $p < 0.10$).

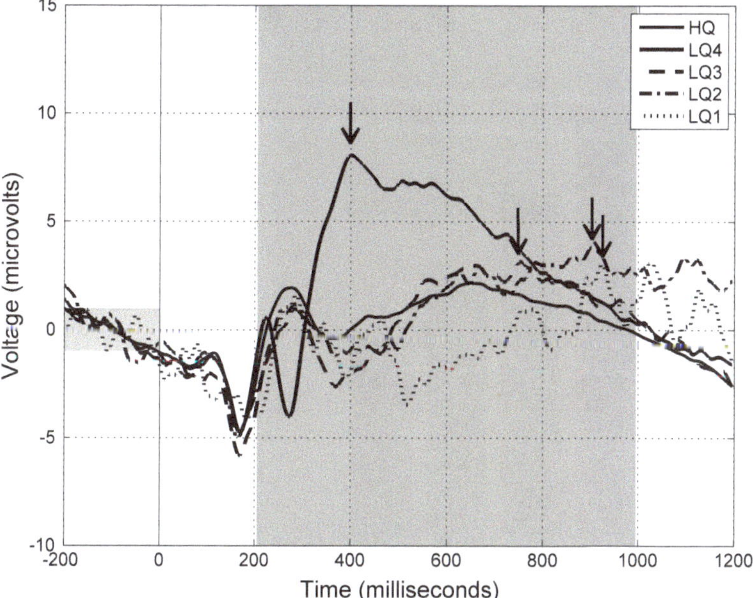

Fig. 2.4 Grand average of ERP plots for HQ and LQ1-4 at channel Cz. For HQ, correctly rejected trials (whereby no quality loss was perceived) and for LQ1-4 hits (whereby a quality loss was perceived) were utilized. *Arrows* denote P300 peaks. Number of trials used for the grand average of the ERP plots per class: HQ = 22,832, LQ4 = 3,268, LQ3 = 1,332, LQ2 = 610, and LQ1 = 165. Reprinted, with permission, from [1]

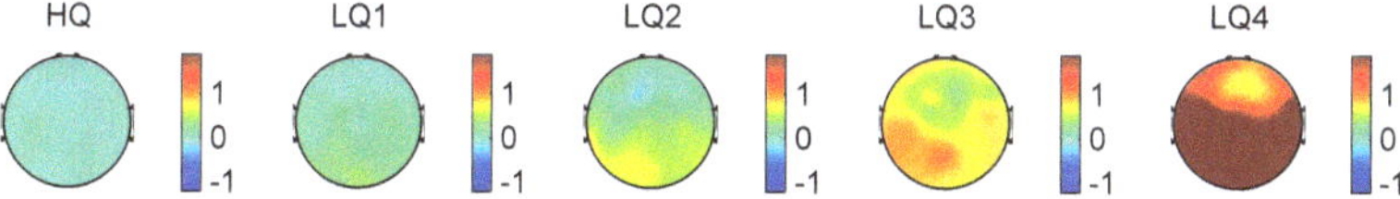

Fig. 2.5 Scalp topographies for all channels. Each *circle* depicts a top view of the head, with the nose pointing upwards. *Colors* code the mean voltage (microvolts) for the time interval from 300 to 1,000 ms after stimulus onset. For LQ1-4, hits were used and for HQ, correctly rejected trials were used. Reprinted, with permission, from [1]

2.5.3 Classification

The classification results can be found in Fig. 2.6. At the first classification level, trained on hits versus HQ and tested on hits versus HQ, the average AUCb value reached a high level for LQ4: AUCb = 0.92 ($p < 0.01$), LQ3: AUCb = 0.85 ($p < 0.01$), LQ2: AUCb = 0.76 ($p < 0.05$), and LQ1: AUCb = 0.70 (ns).

The second classification level reached the following values; for LQ4: not enough misses, LQ3: AUCb = 0.61 ($p < 0.05$), LQ2: AUCb = 0.55 (ns), and LQ1: AUCb = 0.51 (ns).

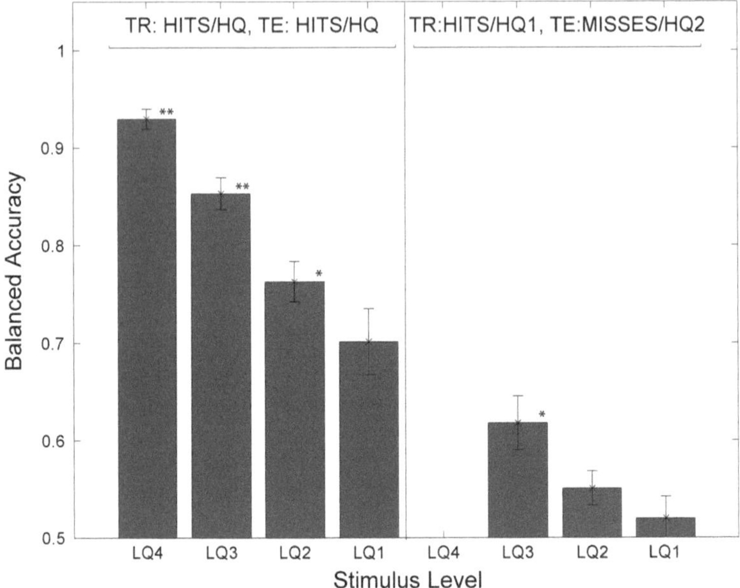

Fig. 2.6 Classification results. *Bars* show the average classification performance (balanced accuracy value). *Left* Trained (TR) on hits versus HQ and tested (TE) on hits versus HQ; *Right* Trained (TR) on hits versus HQ1 and tested (TE) on misses versus HQ2; for all stimuli LQ1-4. The bar for LQ4 is missing, as no participant had enough misses for testing (classification 2). Number of participants used for the average of first classification (*left*): LQ1 4, LQ2 7, LQ3 9, LQ4 10 and for the second classification (*right*): LQ1 = 4, LQ2 = 7, LQ3 = 8, LQ4 = 0. Whiskers denote standard errors. *Asterisks* denote the significance level of the classification outcome in a Wilcoxon rank-sum test (one *asterisk* for $p < 0.05$, and two for $p < 0.01$). Reprinted, with permission, from [1]

It should be noted here that classification could not be carried out for all participants (due to a small number of hit/miss trials), so that the average values reported here are averages calculated over subsets of participants (classification 1: LQ1 = 4, LQ2 = 7, LQ3 = 9, LQ4 = 10; classification 2: LQ1 = 4, LQ2 = 7, LQ3 = 8, LQ4 = 0).

2.6 Discussion

An analysis of the *opinion test* ratings revealed that quality was rated as significantly lower from an SNR below 21 dB and downwards. This point denotes the threshold at which the quality was perceived as significantly worse in comparison to the reference value. The reaction time for the strongest degradation was shorter compared to the weaker ones, implying that participants were faster in detecting the degradation and providing the corresponding rating. The psychometric functions showed that the mean detection rate surpassed 50 % at the 21 dB degradation level. This result is similar to the results of the opinion test.

For the *ERP data*, the significant P300 generated by the control stimulus (i.e. /i/) showed that the experimental set-up was appropriate for its purposes. It might appear surprising that a residual P3 response, which is known to represent cognitive stimulus appraisal, was detected in trials for which no behavioral detection had been reported. In the context of the present paradigm, one could argue that minor physical stimulus differences were initially detected, yet an internal response criterion had not been met, so that an overt behavioral report was not initiated. The effects of *P300 peak latency* discovered here showed that the harder it was to detect a degraded stimulus, the later a P300 was evoked. This could be due to the fact that more cognitive effort is involved in detecting the degradation. The significant variation of the *P300 peak amplitude* is comparable to the variation of latency, but shows the opposite pattern of change: the stronger the degradation, the higher the P300 amplitude. This result was supported by the two correlations.

The P300 amplitude varies with the detection rate: the higher the amplitude, the higher the detection rate. Comparing the amplitude with the latency of the P300, a negative correlation suggests that the smaller the amplitude, the longer the latency.

Interestingly, the analysis of the grand mean data obtained as an average across all participants showed the strongest P3 response at Cz. Thus, in the present paradigm the most effective placement of a single electrode was in-between the commonly reported places for novelty (P3a) and target (P3b) ERPs, which have been described at more frontal or more parietal sites [80]. At the first level of classification, it was demonstrated that the brain reaction due to the processing of a degradation—in this case the difference between the undisturbed and disturbed stimuli—can be well detected. With the second classification, it was shown that the pattern of brain activation related to consciously processed degradations can also be detected in trials which are not reported as degraded on a subjective level. It was concluded that these trials might have been processed non-consciously and had no measurable influence on the direct user rating. This processing might still lead to an influenced long-term quality judgement, due to increased cognitive load and fatigue when exposed to small degradations over a long period of time (for measuring fatigue using EEG see [97]).

2.7 Length Influence Experiment

The developed test set-up cannot be implemented with every possible stimulus length and degradation class. In this remaining part of this chapter, it will be shown how the length of speech stimuli and the headphone type may influence the resulting subjective speech quality ratings. Therefore, three different lengths of stimuli will be analyzed (length of phonemes, words, and sentences).

2.7.1 Introduction

The aim of this experiment was to determine if (1) the length of a stimulus has an influence on the subjective rating and if (2) the type of headphones has an influence on the subjective rating of one degradation class. The motivation for this test was based on the fact that the common length of a stimulus used for tests in telecommunication research is around 8 s [19], which is much longer than the length common in ERP research (between 100 and 1,500 ms). In addition to this, common speech quality tests are carried out with circumaural headphones rather than in-ear headphones, the latter being typical for EEG set-ups. The result will reveal if the headphone type caused an unwanted influence on degradation perception.

2.7.2 Methods

2.7.2.1 Participants

Twenty volunteers (ten female, ten male; average age = 24.32 years; SD = 3.54; range = 22–28 years old; all right-handed), only native German speakers took part in this experiment. All participants reported normal auditory acuity. They provided their informed consent and received monetary compensation.

2.7.2.2 Material

For this experiment, stimuli of three different lengths were used: phonemes, words, and sentences. The phoneme from the first experiment (validation of test set-up, see Chap. 2) was used: /a/ (200 ms). The stimulus with the length of one word was the German translation of "eyebrow" /Augenbraue/ (1,200 ms), and a sentence shortened from the EUROM data base (following [98]) uttered by a male speaker was also used (8,000 ms) [99]. The two tested headphones were Sennheiser in-ear headphones and AKG over-ear headphones. Stimuli were degraded with the use of signal-correlated noise at the following SNRs: 5, 10, 14, 16, 18, 20, 21–35 dB in one-dB increments.

2.7.2.3 Experimental Design and Procedure

As in the first experiment (Chap. 2), participants had to rate all stimuli on a *continuous quality scale* (*CQS*) ranging from bad (0) to excellent (100). The stimuli types (three different lengths) were judged on all levels of degradation (four levels) and with both headphones (in-ear versus over-ear).

2.7.3 Statistical Analysis

The data was analyzed performing an ANOVA for repeated measures with *type of headphone* and *length of stimulus* as the independent variables and the *subjective quality rating* as the dependent variable.

2.7.4 Results

A significant main effect for the factor: *length of stimulus* was found ($F(21{,}4404) = 598.19, p < 0.01, \eta^2 = 0.15$).

As can be seen in Fig. 2.7, ratings for the short phoneme stimulus (average rating = 82.26) were significantly higher compared to the stimuli in word (average

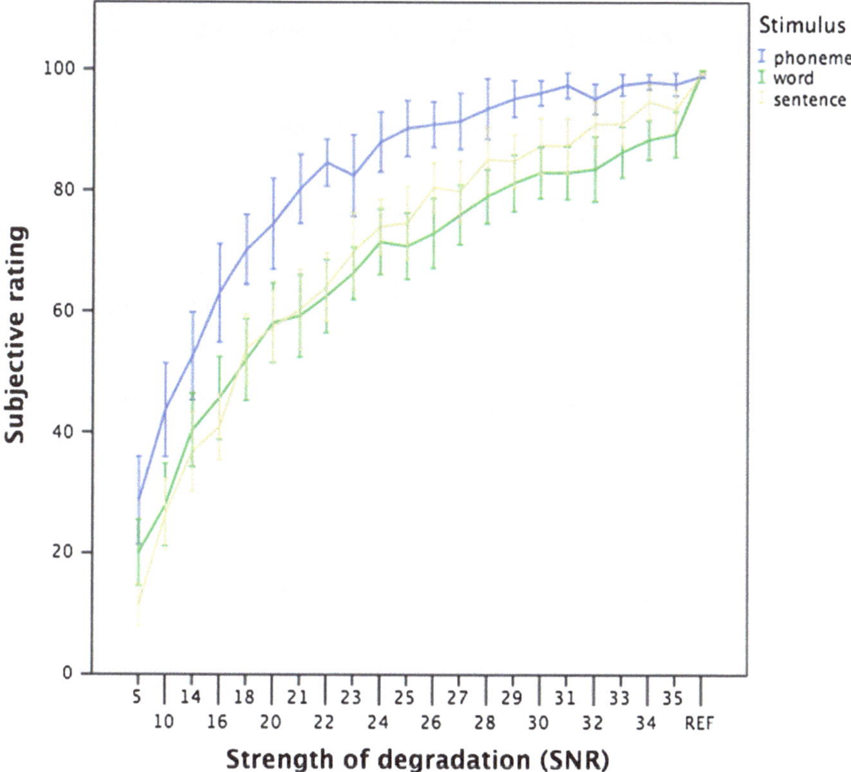

Fig. 2.7 Average rating from the *continuous quality scale* (*CQS*) across all participants. Subjective rating have been displayed as a function of degradation strength. REF denotes the clean reference stimulus. The lengths of stimuli have been color-coded. Whiskers denote the 95 % confidence intervals

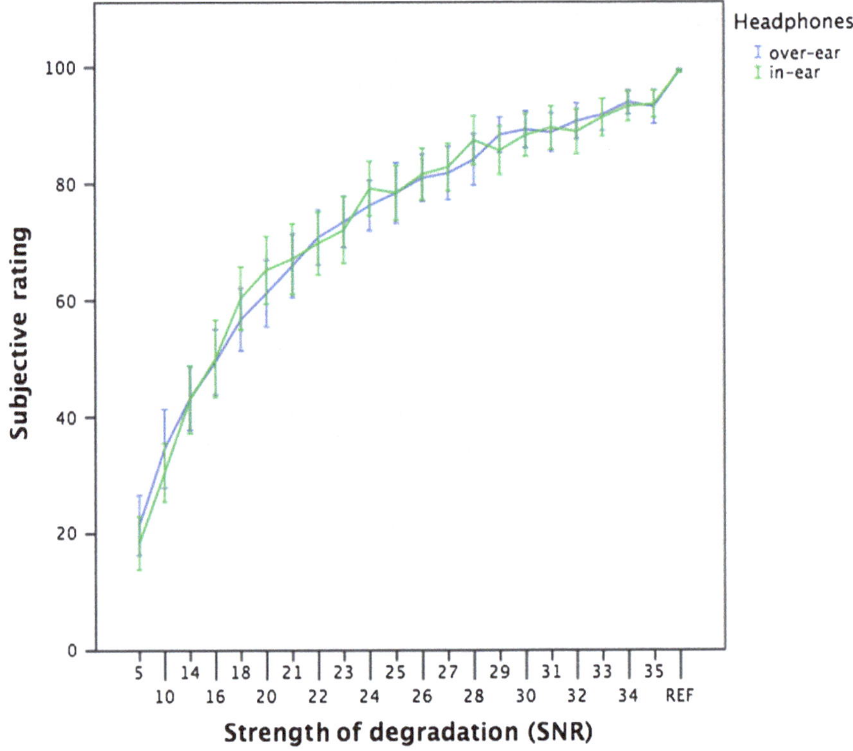

Fig. 2.8 Average rating from the *continuous quality scale* (*CQS*) across all participants. Subjective ratings have been displayed as a function of degradation strength. REF denotes the clean reference stimulus. The type of headphone used has been color-coded. Whiskers denote the 95 % confidence intervals

rating = 68.00) and sentence length (average rating = 70.13). There was no significant difference between stimuli with differing lengths of words and sentences.

There was no main effect for the *type of headphone* ($F(1,4404) = 0.58$, ns). As can be seen in Fig. 2.8, the confidence intervals overlap for stimuli played with in-ear (average rating = 73.52) and over-ear headphones (average rating = 73.41).

The post-hoc analysis (Sidak adjustment for pairwise comparisons) revealed a significant difference between the stimuli associated with the length of phonemes and words ($p < 0.01$), as well as for the difference between phonemes and sentences ($p < 0.01$). The difference between stimuli associated with the length of words and sentences was not significant.

2.7.5 Discussion

As expected from the *Phonemes Experiment* (Chap. 2), the *level of degradation* had an influence on subjective judgement. The factor *length of stimulus* had a significant effect on the subjective quality rating. A significantly higher quality was assigned to stimuli associated with the length of phonemes compared to stimuli associated with the length of words and sentences. There was no difference between the judgment of word-long and sentence-long stimuli. This leads to the conclusion that stimuli for subjective experiments on quality should consider stimuli minimum of word length. For EEG experiments concerning quality, it can be argued that stimuli should have word length as well. It still remains unclear whether the sensitivity of physiological measurement is higher, and therefore, here shorter stimuli can also possibly be valid for testing. As there was no difference between the ratings of *the two headphone types*, the influence may be negligible.

The *Phonemes Experiment* (Chap. 2) used short stimuli (vowels) which are a standard in ERP studies and this allowed the use of established physiological knowledge to interpret new findings and their implications for the cerebral processing of stimulus quality. However, in quality research, longer stimuli are employed for the behavioral detection of stimulus degradation. Correspondingly, the *Length Influence Experiment* directly compared the effects of stimuli differing in length (vowels, words, sentences) on the perceived quality in an opinion test. Indeed, longer stimuli (i.e., words or sentences) permit the better detection of minor stimulus degradation. In response to this behavioral result, the second combined experiment (the *Word Experiment*) will be introduced in the next chapter (Chap. 3), which makes use of word stimuli, thereby linking the ERP results presented here directly to the subjective standards in quality research.

2.8 Chapter Summary

In this chapter, a combined test set-up was introduced. During the stimulation of short speech stimuli (phonemes) of varying quality, EEG signals were measured. In addition to this, an opinion test was carried out. The results show that the physiological response, measured as parameters of an ERP component, can be used to gain insight into the perceived stimulus material. The P300 parameters vary with degradation strength, and therefore, can probably be used to estimate the quality of the presented stimulus material. As a second major contribution, the test set-up was validated using short speech stimuli and signal-correlated noise as degradation. This result is only true for the average response across a group of participants and can vary if applied to single participants. For the comparison with standard subjective tests, averaging methods using subjective, instrumental, and then physiological variables is a valid approach when it comes to standardization and technological developments intended for larger groups of listeners.

Chapter 3
ERPs and Quality Ratings Evoked by Word Stimuli and Varying Bit Rate Conditions

In the previous chapter (Chap. 2), it was shown that the length of speech stimuli has an influence on the resulting subjective speech quality ratings. Three different lengths of stimuli were tested (length of phonemes, words, and sentences). The results established that longer stimuli (i.e., words or sentences) permit the better detection of minor stimulus degradation. In response to this subjective result, the second combined experiment (*Word Experiment*) described in this chapter will investigate whether the combined EEG/opinion test set-up also functions with word-long stimuli and differing bit rate conditions of a speech codec as degradation factors (for an overview see Fig. 3.1).

This is the third major contribution of this book, showing that the test set-up developed in this context can be applied to longer stimuli and other degradation classes.

3.1 Introduction

The objective of the *Word Experiment* was to test even more realistic stimuli in terms of their length and to extend the test set-up to include another class of degradation.[1] The stimuli were words (/Haus/ (English: house) and /Schild/ (English: sign)) and each differed with regard to the bit rate of the codec according to the ITU-T Recommendation G.722.2 [7]. The selection of these two words was based on fact the test vowels should consistently—in line with the *Phoneme Experiment*—contain the vowels /a/ and /i/. Furthermore, the selection was carried out by selecting short words with varying consonant intonation, with clear articulation, and high energy expenditure. The difference between the high-quality (HQ; wideband) and lower qualities

[1] This chapter is based on a previous publication; text fragments, tables, and figures are based on Antons et al. 2012a [1]. Reprinted, with permission, from [1].

J.-N. Antons, *Neural Correlates of Quality Perception for Complex Speech Signals*, T-Labs Series in Telecommunication Services, DOI 10.1007/978-3-319-15521-0_3

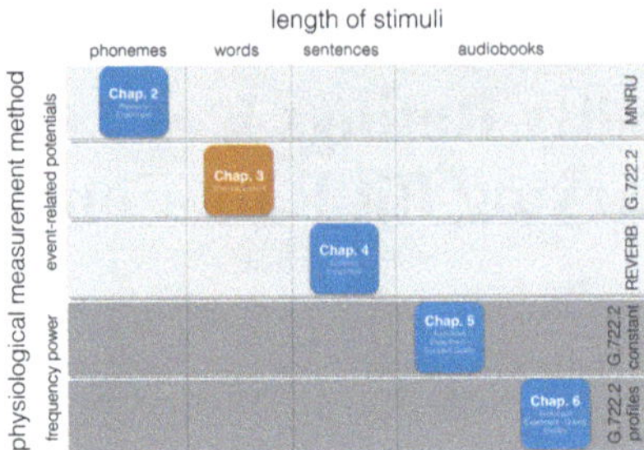

Fig. 3.1 Conducted experiments and structure of this book. Different lengths of stimuli on the x-axis (phonemes, words, sentences, and audiobooks). Physiological measurement techniques on the y-axis (EEG frequency band power and event-related potentials). Applied classes of degradations are color-coded (*grey bars*) and indicated on the *right* (signal-correlated noise introduced by a modulated noise regulation unit (MNRU), bit rate reductions introduced by using different settings of a speech codec in accordance with ITU-T Rec. G.722.2, and reverberation (REVERB) introduced by different room impulse responses. Current chapter is indicated in *orange*. Chapter 3, *Word Experiment*

(LQ1-4; subset chosen from the conditions: 6.6, 8.85, 12.65, 14.25, 15.85, 18.25, 19.85, 23.05 kbit/s) was expected to elicit an early difference pattern for conditions LQ2-4 as well as a P300 for at least the highest degradation.

3.2 Methods

3.2.1 Participants

Nine participants (four female, five male; average age = 25.22 years; SD = 1.20; range = 24–27 years old; all right-handed), only native German speakers took part in the *Word Experiment*. None of them had participated in the *Phoneme Experiment* (Chap. 2). All participants reported normal auditory acuity. They provided their informed consent and received monetary compensation.

3.2.2 Material

A different stimulus material than in the *Phonemes Experiment* were chosen. Two words were spoken by a female and a male speaker. For all four LQ conditions, the codec G.722.2 was used as a degradation factor. The best quality was used as high quality (HQ: direct wideband quality without any coding-decoding process) as the standard stimulus, while all lower bit rate conditions were used as deviants (6.6, 8.85, 12.65, 14.25, 15.85, 18.25, 19.85 and 23.05 kbit/s). All combinations resulted in a stimulus set of nine per word and speaker. As in the *Phonemes Experiment* (Chap. 2), a subset of four stimuli was determined individually for each participant.

Table 3.1 Bit rate (kbit/s) for all participants and medians

Participant	kbit/s				
	HQ	LQ1	LQ2	LQ3	LQ4
1	WB	14.25	12.65	8.85	6.6
2	WB	14.25	12.65	8.85	6.6
3	WB	14.25	12.65	8.85	6.6
4	WB	15.85	14.25	12.65	8.85
5	WB	23.05	19.85	18.25	15.85
6	WB	15.85	14.25	12.65	8.85
7	WB	15.85	14.25	12.65	8.85
8	WB	14.25	12.65	8.85	6.6
9	WB	14.25	12.65	8.85	6.6
Median	WB	14.25	12.65	12.65	6.6

During the pre-test, all stimuli were presented to the participants. The task of the participants was to rate a stimulus as belonging to one of two classes, and as having high quality (no degradation) or low quality (with degradation). A detection rate was calculated by dividing the number of correctly identified low quality stimuli by the total number of presented stimuli in the corresponding category. Based on the resulting detection rate, a final stimulus selection consisting of four stimuli was carried out. The targeted detection rates were LQ4 = 100 %, LQ3 = 60 %, LQ2 = 40 %, and LQ1 = 0 %. The selected stimulus levels can be found in Table 3.1.

3.2.3 Experimental Design and Procedure

In a forced choice task, participants had to rate whether a given word was of high quality (HQ) or degraded (LQ). Stimuli were presented either in high quality or were impaired. Besides the aforementioned modifications, the same experimental settings as in the *Phonemes Experiment* (Chap. 2) were utilized. Participants had to rate all stimuli on a *continuous quality scale* (*CQS*) ranging from bad (0) to excellent (100). For this test, the *MUlti Stimulus test with Hidden Reference and Anchor* (*MUSHRA*) was adapted to match as closely as possible the EEG experiment that was carried out. Therefore, no hidden reference and no anchor stimulus were used.

3.2.4 Electrophysiological Recordings

Settings for the electrophysiological recordings were the same as in the *Phonemes Experiment* (Chap. 2). Scalp locations: Fp1-2; AF3-4; Fz, 1-6, 9-10; FCz, 1-8; T7-8; Cz, 1-6; TP7-8; CPz, 1-6; Pz, 1-10; Poz, 3-4, 7-8; Oz, 1-2; AF7-8 and the right mastoid were recorded.

3.2.5 Data Analysis

The data was analyzed in the same way as in the *Phonemes Experiment* (Chap. 2) except the following changes. The time frame for P300 quantification was set at 400–900 ms after stimulus onset. As time frames for classification, 400–500, 500–600, 600–700 and 700–900 ms were used.

3.3 Statistical Analysis

Statistical analysis was performed in the same way described for the *Phonemes Experiment* (Chap. 2).
Behavioral data: An analysis of variance (ANOVA)—with *degradation intensity* as the independent variable and the *mean opinion score* (*MOS*) as the dependent variable—was calculated.
ERP data: The *peak latency* and *peak amplitude* of the *P300* responses were analyzed by means of repeated measures ANOVA.
Classification: A *Linear Discriminant Analysis* (*LDA*) was used for the classification of the extracted P300 peak amplitude and latency.

3.4 Results

3.4.1 Behavioral Data

The ANOVA calculated on the subjective data, with *degradation intensity* as the independent variable and the *mean opinion score* (*MOS*) as the dependent variable, revealed a significant influence on the factor: *degradation intensity* ($F(120, 4262) = 550.86$, $p < 0.05$, $\eta^2 = 0.64$). The posthoc test (Sidak adjustment for pairwise comparisons) reached significance at a level of 8.85 kbit/s ($p < 0.05$). The mean reaction times for the different conditions can be found in Table 3.2.

The reaction times for LQ1-3 are on a similar level, but significantly different compared to LQ4 ($p < 0.05$) and HQ ($p < 0.05$). For the stimulus condition LQ4 and HQ, reaction times were shorter. The psychometric function fits for all participants have been plotted in Fig. 3.2.

Table 3.2 Mean reaction times for all conditions

	Milliseconds				
	HQ	LQ1	LQ2	LQ3	LQ4
Mean	609.17	724.93	736.28	749.24	598.54

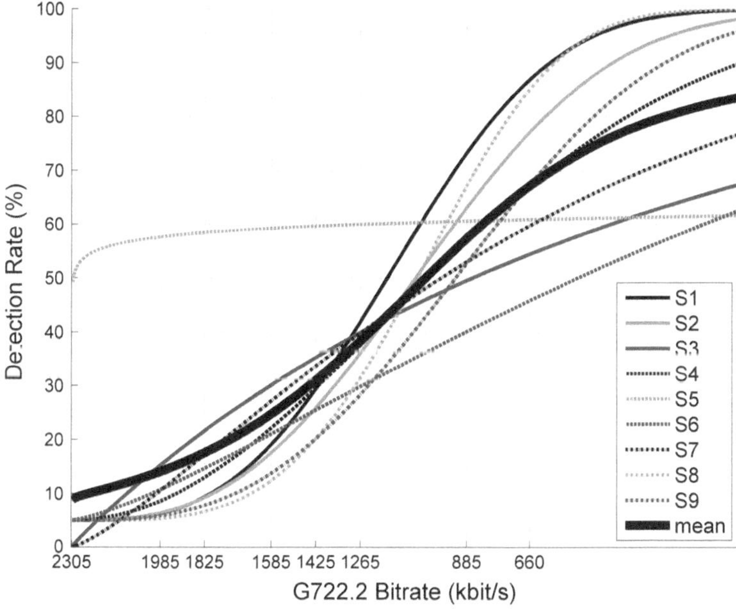

Fig. 3.2 Psychometric function fits of the psychophysical data; for all participants (participant(S) 1–10) and the mean across all participants. Reprinted, with permission, from [1]

3.4.2 P300

The ANOVA for repeated measurements revealed a main effect on the factor: *degradation intensity* ($F(4, 3) = 3.36, p < 0.01, \eta^2 = 0.35$). The *P300 peak amplitude* variation is significant ($F(3, 18) = 10.10$, $p < 0.01$, $\eta^2 = 0.62$). For the dependent variable P300, *peak latency*, no effect was found. The grand average for all stimulus classes can be found in Fig. 3.3.

The pairwise comparison for the *peak amplitude* revealed a significant difference between LQ1 and LQ2 ($p < 0.05$) as well as LQ1 and LQ4 ($p < 0.05$). Figure 3.4 shows the scalp distribution of voltage for the different stimulus conditions (hits and correct rejections for LQ1-4 and HQ, respectively). For LQ4, a broad reaction was detected. For the less disturbed stimuli, a reaction was evoked, though not as strong as for LQ4.

3.4.3 Classification

The results for the classification can be found in Fig. 3.5.

At the first classification level, trained on hits versus HQ and tested on hits versus HQ, the average AUCb value reached a high level (for LQ1: AUCb = 0.72 ($p < 0.01$),

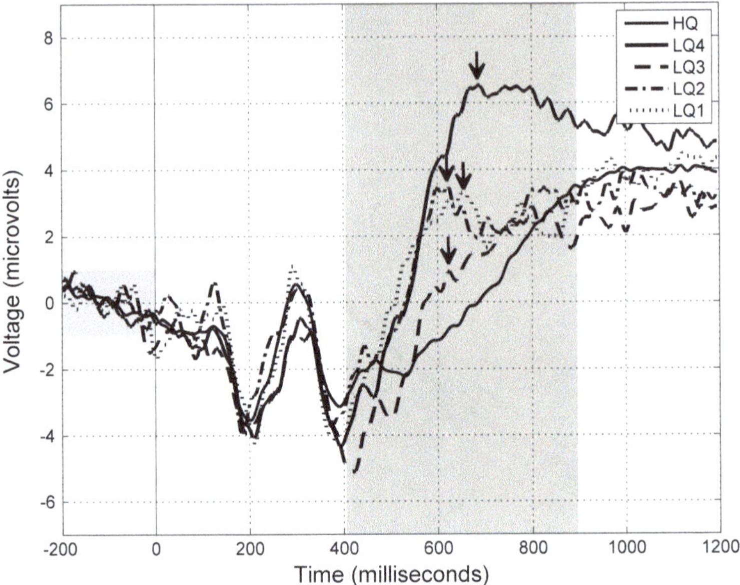

Fig. 3.3 Grand average ERP plots for HQ and LQ1-4 at channel Cz. For HQ correctly rejected trials (whereby no quality loss was perceived) and for LQ1-4 hits (whereby quality loss was perceived) were used. *Arrows* denote P300 peak. Number of trials used for the grand average of ERP plots per class: HQ = 11177, LQ4 = 1235, LQ3 = 826, LQ2 = 655 and LQ1 = 500. Reprinted, with permission, from [1]

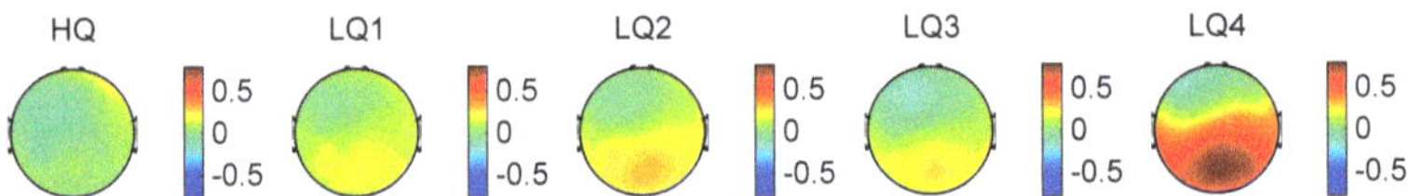

Fig. 3.4 Scalp topographies for all channels. Each *circle* depicts a top view of the head, with the nose pointing upwards. Colors code the mean voltage (microvolts) for the time interval from 500 to 1,000 ms after stimulus onset. For LQ1-4, hits and for HQ, correctly rejected trials were used. Reprinted, with permission, from [1]

LQ2: AUCb = 0.66 ($p < 0.01$), LQ3: AUCb = 0.62 ($p < 0.01$), and LQ4: AUCb = 0.59 ($p < 0.01$)). The second classification level reached the following values; for LQ1: AUCb = 0.53 ($p < 0.05$), LQ2: AUCb = 0.54 (ns), LQ3: AUCb = 0.53 ($p < 0.05$), and LQ4: AUCb = 0.53 (ns).

As for the *Phoneme Experiment*, it should be considered that the average values reported here are mean values calculated across subsets of participants (classification 2: LQ4: 7 participants; otherwise: all participants).

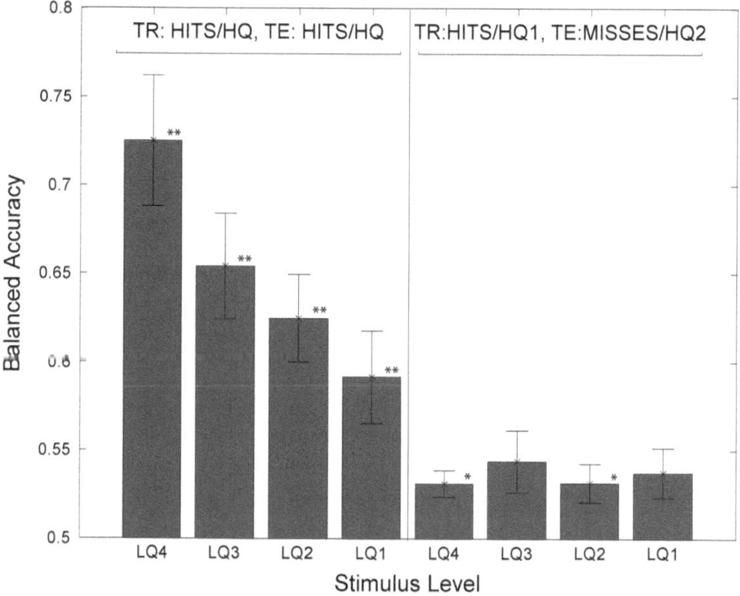

Fig. 3.5 Classification results. *Bars* show the average classification performance (balanced accuracy value). *Left* Trained (TR) on hits versus HQ and tested (TE) on hits versus HQ; *Right* Trained (TR) on hits against HQ1 and tested (TE) on misses against HQ2; for all stimuli LQ1-4. Number of participants used for the average of first classification (*left*): LQ1 = 9, LQ2 = 9, LQ3 = 9, LQ4 = 9 and for the second classification (*right*): LQ1 = 9, LQ2 = 9, LQ3 = 9, LQ4 = 7. Whiskers denote standard errors. *Asterisks* denote the significance level of the classification outcome in a Wilcoxon rank-sum test (*one asterisk* for $p < 0.05$, and two for $p < 0.01$). Reprinted, with permission, from [1]

3.5 Discussion

An analysis of the *opinion test* ratings revealed that the quality was rated as significant lower than a bit rate of 8.85 kbit/s downwards.

Surprisingly, one of the participants was more sensitive concerning the detection of the degradation factor (Fig. 3.2). Even after a detailed inspection of the data, no irregularity of the ERP data was encountered, and thus, the data was finally included in the analysis.

The *reaction time* for the strongest degradation and for the HQ stimulus was lower compared to the more weakly degraded ones, indicating that participants were faster to detect the degradation factor and to provide the corresponding rating. It is perhaps surprising that the reaction time for the best and lowest quality levels were shorter. This effect is due to the fact that reaction time is related to task difficulty. It appears that it was easier to detect the extremes (no or much degradation) in contrast to the medium degraded stimulus level.

The *psychometric functions* showed that the mean detection rate surpassed 50 % at the 8.85 kbit/s degradation level. Compared with the psychometric functions of the *Phonemes Experiment* (Chap. 2), the curves of the *Words Experiment* (Chap. 3) are smoother, indicating that the detection rate rose slower with the intensity of the degradation. This is due to the fact that participants can clearly identify the noise in the *Phoneme Experiment* as a degradation, whereas the compression artifacts in this experiment were harder to detect for some participants.

The significant variation of the P300 mean amplitude is comparable to the variation registered in the *Phoneme Experiment*: the stronger the degradation, the higher the P300 mean amplitude.

At the *first level of classification*, it was demonstrated that the brain reaction associated with the processing of a degradation, in this case the difference between the undisturbed and disturbed stimulus, can be well detected. With the *second classification*, it was shown that the pattern of brain activity related to the conscious processing of degradations can also be detected in trials which are not reported as degraded on a subjective level. It must be concluded once again that these trials might have been processed non-consciously and had no measurable influence on the direct user rating. Balanced accuracy for second classification shows a remarkable similarity across quality levels, in contrast to what might have been expected. This might be the case, because one stimulus had a surprisingly high number of hits at low quality levels for a considerable number of participants (LQ2 and 1).

3.6 Chapter Summary

In this chapter (*Word Experiment*) it was shown that the combined EEG/opinion test set-up is also valid for word-long stimuli and differing bit rate conditions of a speech codec as a degradation factor. This is the third major contribution of this book, showing that the test set-up developed in this context can be applied to longer stimuli and other degradation classes.

In the next chapter (Chap. 4, *Sentence Experiment*), the generalization of the test set-up will be advanced further and applied to sentence-long stimuli and reverberation as a degradation factor.

Chapter 4
ERPs and Quality Ratings Evoked by Sentence Stimuli at Different Reverberation Levels

In the previous chapter (Chap. 3), the second experiment (*Word Experiment*) showed that the combined EEG/opinion test set-up is also valid for word-long stimuli and differing bit rate conditions of a speech codec as degradation factors.

In this chapter (Chap. 4, *Sentence Experiment*), the generalization of the test set-up will be further applied to sentence-long stimuli at different reverberation levels. Therefore, a speech stimulus with the length of one sentence will be presented and—as degradation factors—different reverberation conditions will be employed (for an overview see Fig. 4.1).

4.1 Introduction

This chapter aims to investigate the use of EEG to identify neural and affective correlates of speech quality perception at *different reverberation levels*.[1] Focus has been placed on hands-free speech communications where reverberation can severely degrade the signal timbre [100, 101], cause temporal smearing, and ultimately degrade speech quality and intelligibility [102]. Here, two reverberant environments have been considered: a domestic living room and a large auditorium.

4.2 Methods

4.2.1 Participants

Twenty-two participants took part in the *Sentence Experiment* (ten female, twelve male; average age = 23.40 years; SD = 3.80; range = 18–33 years old); all of them were fluent English speakers. Please note that this experiment was performed in the

[1] This chapter is based on a previous publication; text fragments, tables, and figures are based on Antons et al. [3]. Reprinted, with permission, from [3].

J.-N. Antons, *Neural Correlates of Quality Perception for Complex Speech Signals*, T-Labs Series in Telecommunication Services, DOI 10.1007/978-3-319-15521-0_4

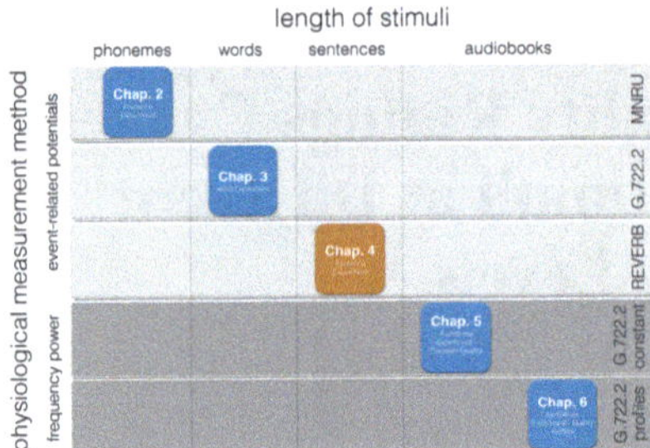

Fig. 4.1 Conducted experiments and structure of this book. Different lengths of stimuli on the x-axis (phonemes, words, sentences, and audiobooks). Physiological measurement techniques on the y-axis (EEG frequency band power and event-related potentials). Applied classes of degradations are color-coded (*grey bars*) and indicated on the *right* (signal-correlated noise introduced by a modulated noise regulation unit (MNRU), bit rate reductions introduced by using different settings of a speech codec in accordance with ITU-T Rec. G.722.2, and reverberation (REVERB) introduced by different room impulse responses. Current chapter is indicated in *orange*. Chapter 4, *Sentence Experiment*

English language, therefore, listeners who were also fluent in English were recruited. All participants reported normal auditory acuity and no medical problems. Participants provided their informed consent and received monetary compensation for their participation. The experimental protocol was approved by the Research Ethics Office at INRS-EMT and at McGill University (Montreal, Canada).

4.2.2 Material

As test stimulus, a double-sentence utterance common in subjective quality tests was used. The sentence was uttered by a male speaker in an anechoic chamber and digitized at an 8 kHz sampling rate with 16-bit resolution. Room impulse responses recorded in a typical domestic living room environment (reverberation time of 400 ms) and in an auditorium (reverberation time of 1,500 ms) were convolved with the clean speech file to generate the reverberant stimuli. To ensure consistency, all files were normalized to −26 dBov using the ITU-T P.56 voltmeter [103]. Unlike typical subjective quality tests, only one speech file (three stimuli: one clean and two reverberant) was used here in order to maintain a controlled content, as P300 signals can be sensitive to varying content.

4.2.3 Experimental Design

The experimental protocol in this chapter contained two parts. The first consisted of a quantitative "pre-test" component where participants (i) filled in a demographic questionnaire, (ii) performed a subjective quality test using the *Absolute Category*

Rating (*ACR*) scale [19] (5-point scale with 1 indicating bad quality and 5 excellent), and (iii) rated their elicited emotional states after hearing the different speech files. In contrast to the rating task in Chaps. 2 and 3 where participants rated on a continuous quality scale, absolute rating was performed by the participants in this experiment. This is due to the fact that the length of the stimuli is long enough to be validly judged without having an available reference stimulus. Furthermore, in this experiment participants were asked to rate their arousal and valence levels. This emotional self-assessment could be an indicator of how quality influences (1) the cognitive state (arousal) and the liking (valence). In order to evaluate the emotional self-assessment, modified versions of the *Self-Assessment Manikin* (*SAM*) scales were used [104] (see a more detailed description of the SAM scales in Sect. 1.2). More specifically, listeners rated the arousal, valence, and dominance dimensions using 9-point visual anchors. Lastly, in order to gauge the participants "experience" with the test, they were also asked to rate their "liking" using a 9-point scale (1 (not at all) to 9 (very much)) and how familiar they are with the type of degradation using a 5-point scale (1 (not at all) to 5 (very much)).

The selection of subjective methods for this experiment is based on the fact that—besides the overall quality rating (ACR)—the emotional state of participants can also vary due to stimulus parameter, e.g., if the quality of a transmitted speech stimulus is low, participants could get annoyed or angry. Therefore, the SAM was included in the testing procedure. The liking scale was included, as it would show whether this scale was similar to the overall quality rating and the rating on the SAM scale valence. In addition, participants were asked if they were familiar with the degradation presented to them, which was indented to show whether the overall quality rating (based previous experiences with this class of degradations) would influence the result.

During the pre-test, participants listened to each speech file (three stimulus level) three times in random order.

The second part of the test consisted of the actual EEG experiment in accordance with an oddball paradigm (see Sect. 1.3.4.5). More specifically, the clean speech file served as the so-called *standard* stimulus (70 % of the trials) and the reverberant files served as *deviants* (30 % of the trials). Clean and reverberant speech files were delivered in a pseudo-randomized order, forcing at least one standard to be presented between successive deviants, in sequences of 100 trials. Stimulus sequences were presented with an interstimulus interval varying from 1,000 to 1,800 ms. Participants were seated comfortably and were instructed to press a button to indicate whether they detected the clean stimulus or one of the deviants. Stimuli were presented binaurally through in-ear headphones at the listening level preferred by the individual.

4.2.4 Electrophysiological Recordings

A 128-channel BioSemi EEG system was used, but only the following subset of electrodes was utilized for recording: 64 EEG-electrodes according to the *10–20 system* (AFz, 3-4, 7-8; Cz, 1-6; CPz, 1-6; Fz, 1-8 FCz, 1-6; Fpz, 1-2; FT7-8; Iz; Oz, 1-2; Pz, 1-10; POz, 3-4, 7-8; T7-8; TP7-8), 4 EOG-electrodes, and two mastoid electrodes (right and left). Data was recorded at 512 Hz but down-sampled to 200 Hz and bandpass-filtered between 1 and 40 Hz for offline analysis. All channels were re-referenced to the average of all EEG-channels. EEG epochs with a length of 2,700 ms, time-locked to the onset of the stimuli, including a 600 ms pre-stimulus baseline, were extracted and averaged separately for each stimulus level and for each participant. In order to quantify the deviance-related effects of P300, the peak amplitude was measured at electrode Cz within a fixed time frame relative to the pre-stimulus baseline. The time frame for P300 quantification was set at 200–600 ms after stimulus onset. The maximal positive amplitude in this time frame was automatically determined and its voltages were derived for further analysis. Reaction time was also computed for each presented stimulus and consisted of the time between stimulus onset and the actual pressing of the button.

4.3 Results

In order to analyze the data a repeated measurement analysis of variance was carried out using the independent variable *level of reverberation* as well as the dependent variables *MOS, valence, arousal, dominance, P300 peak amplitude, and reaction time*. For the analysis of *liking* and *familiarity*, a Wilcoxon test for paired data ranks was used. In the following subsections, test results for the main effects and the Scheffé-adjusted post-hoc comparisons will be reported. Additionally, correlations (Pearson and Spearman) between quantitative parameters and EEG features will be reported.

4.3.1 Behavioral and Subjective Data

For the *MOS* parameter, a significant main effect for reverberation level ($F(2,16) = 128.89, p < 0.01, \eta^2 = 0.94$) was observed. The plot in Fig. 4.2 depicts the subjective MOS versus the reverberation time curve obtained.

As can be seen in Fig. 4.2, a monotonic decrease in MOS was observed as reverberation time increased. Fig. 4.3, in turn, depicts the three emotional SAM dimensions (arousal, valence, and dominance) versus reverberation time. The *arousal* dimension achieved a main effect for *reverberation* ($F(2,16) = 5.45, p < 0.05, \eta^2 = 0.40$), whereas a significant main effect was found for the dimension *valence* ($F(2,16) = 91.85, p < 0.01, \eta^2 = 0.86$) and *dominance* ($F(2,16) = 9.00, p < 0.01, \eta^2 = 0.52$). As can be seen in Fig. 4.3), a monotonic decrease across all three emotion dimensions was observed with increased reverberation time.

Moreover, significant main effects were also observed for the *liking* ($F(2,16) = 45.88$, $p < 0.01$, $\eta^2 = 0.85$) and *familiarity* experience scales ($F(2,16) = 22.07$,

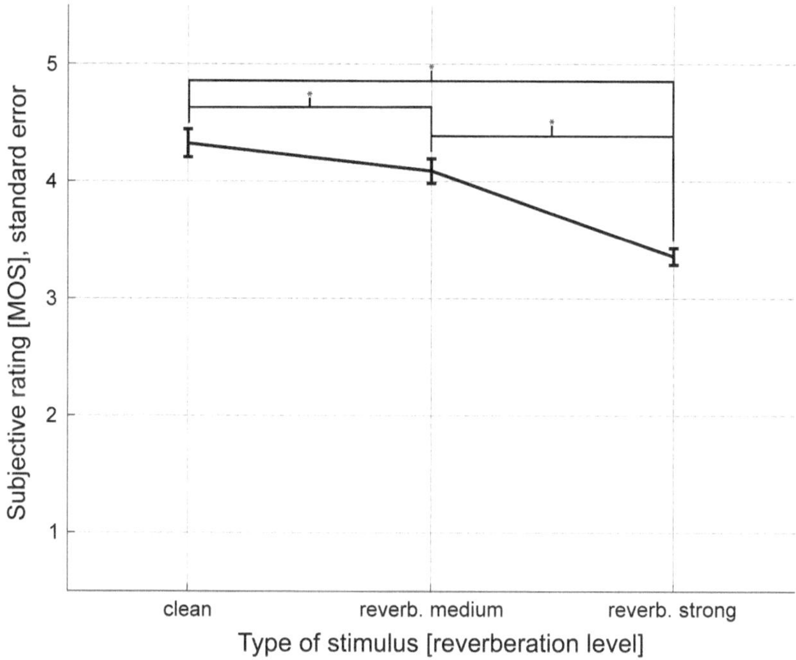

Fig. 4.2 Plots of subjective MOS versus reverberation levels averaged across all participants. Whiskers denote standard errors. *Asterisks* denote the significance level of the pairwise post-hoc comparison ($p < 0.05$). Labels correspond to: reverb. medium = reverberation time = 400 ms and reverb. strong = reverberation time = 1,500 ms. Reprinted, with permission, from [3]

$p < 0.01$, $\eta^2 = 0.73$); plots have been omitted, as monotonically decreasing curves were also observed with increased *reverberation time*, thus, plots would show the same trend as Fig. 4.3. The Wilcoxon test also showed significant effects for both parameters. Results of the post-hoc comparisons have been reported in Table 4.1.

As can be observed, significant differences were identified for all parameters for the clean versus reverberation time (RT) = 400 ms scenario (column labelled "RT0 vs. RT1" in the Table) and for the clean versus RT = 1,500 ms ("RT0 vs. RT2") scenario, with the exception of the arousal dimension. In the RT = 400 ms versus RT = 1,500 ms ("RT1 vs. RT2") scenario, only the MOS, valence, and liking scales were significantly different. Lastly, Table 4.2 reports the correlation matrix for all the collected subjective parameters.

As can be seen in Table 4.1 the *dominance* dimension is only significantly correlated with the *valence* dimension.

Particularly interesting are the high correlations obtained between *MOS* and *valence*, *MOS* and *liking* and *valence* and *liking*, thus indicating that affective states, quality perception, and Quality of Experience (QoE) are inter-related parameters. This suggests that the higher the MOS ratings are, the higher the valence and ratings. *Liking* and *valence* are also strongly correlated, which is not surprising, as the two concepts strongly overlap.

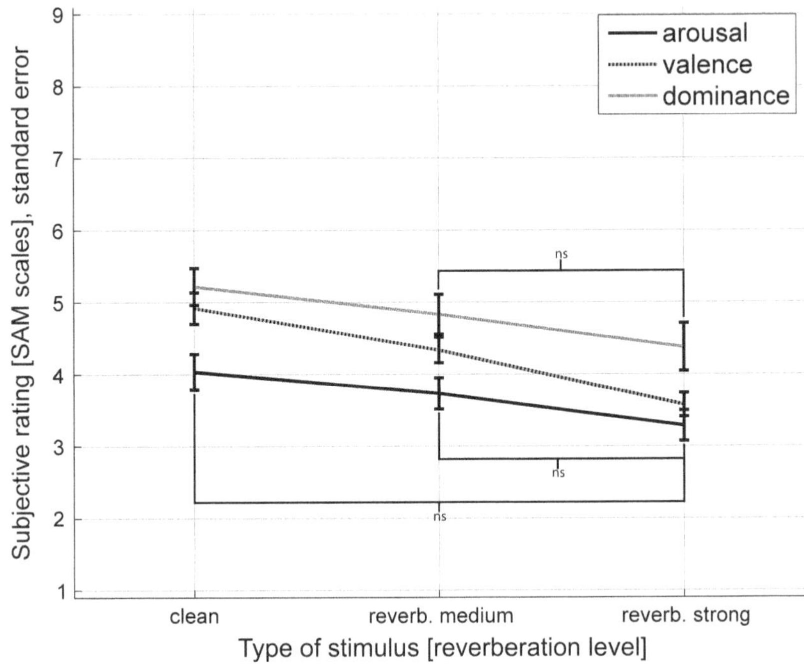

Fig. 4.3 Plots of SAM (*arousal, valence, and dominance*) versus reverberation levels averaged across all participants. Whiskers denote standard errors. Asterisks denote the non-significant pair-wise post-hoc comparisons, all other comparisons are significant ($p < 0.05$). Labels correspond to: reverb. medium = reverberation time = 400 ms and reverb. strong = reverberation time = 1,500 ms. Reprinted, with permission, from [3]

Table 4.1 Scheffé-adjusted post-hoc comparisons

Parameter	RT0 versus RT1	RT0 versus RT2	RT1 versus RT2
MOS	$p < 0.05$	$p < 0.05$	$p < 0.05$
Arousal	$p < 0.05$	ns	ns
Valence	$p < 0.05$	$p < 0.05$	$p < 0.05$
Dominance	$p < 0.05$	$p < 0.05$	ns
Liking	$p < 0.05$	$p < 0.05$	$p < 0.05$
Familiarity	$p < 0.05$	$p < 0.05$	ns

Column labels correspond to: RT0 versus RT1 = clean versus RT = 400 ms; RT0 versus RT2 = clean versus RT = 1,500 ms; RT1 versus RT2 = RT = 400 ms versus RT = 1,500 ms; *ns* not significant

Lastly, it was observed that the *response time* versus *reverberation time* curve was non-monotonic. More specifically, the average response times across all participants were 604, 739, and 691 ms for the clean, RT = 400 ms and RT = 1,500 ms stimuli, respectively.

Table 4.2 Correlation matrix of different quantitative parameters

Parameter	A	V	D	L	F
MOS	0.43**	0.81**	0.19	0.71*	0.37**
Arousal (A)	1	0.65**	0.14	0.38**	0.41**
Valence (V)	–	1	0.34*	0.79**	0.51**
Dominance (D)	–	–	1	0.43**	0.13
Liking (L)	–	–	–	1	0.57**
Familiarity (F)	–	–	–	–	1

** $p < 0.01$ and * $p < 0.05$

4.3.2 P300

Lastly, the neural/cognitive correlates of speech quality perception were investigated. Fig. 4.4 shows the grand average ERPs, the arrows indicate the location of the *P300 peak* for each reverberation level. It was observed that a significant main effect was present for the *P300 peak amplitude* versus *reverberation time* ($F(2,16) = 8.15$,

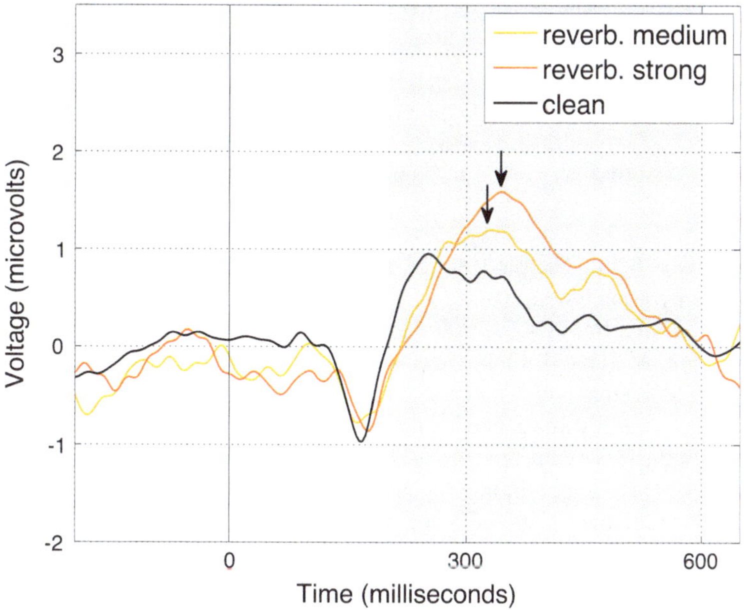

Fig. 4.4 Exemplary grand average of ERP plots for the clean stimulus and reverberation levels (*medium and strong*) at channel CPz. For the clean stimulus, correctly rejected trials (whereby no quality loss was perceived) and for reverberation levels hits (whereby a quality loss was perceived) were utilized. *Arrows* denote P300 peaks. Number of trials used for the grand average of the ERP plots per class: clean = 14,449, reverb. medium = 1,928, reverb. strong = 1,926. Labels correspond to: reverb. medium = reverberation time = 400 ms and reverb. strong = reverberation time = 1,500 ms

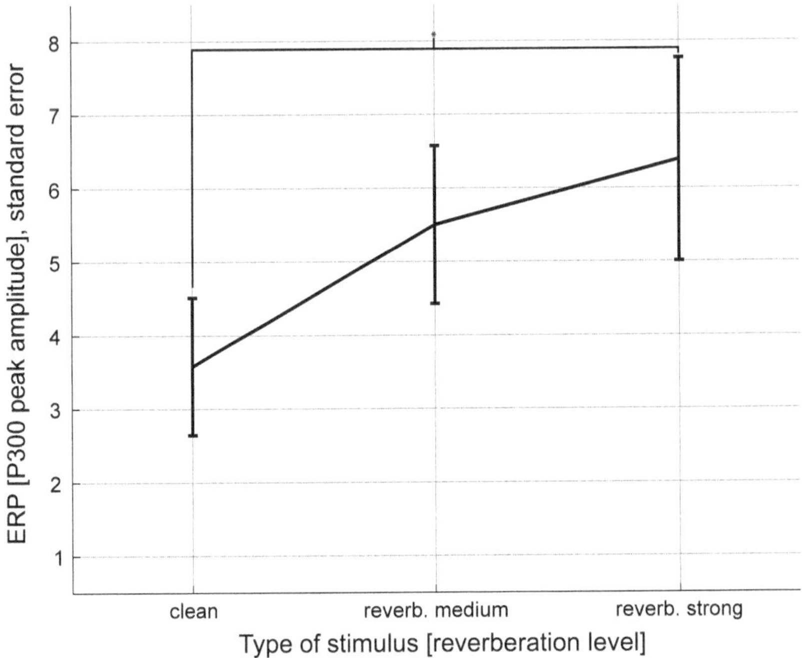

Fig. 4.5 Plots of P300 peak amplitudes versus reverberation levels averaged across all participants. Whiskers denote standard errors. *Asterisks* denote the significance level of the pairwise post-hoc comparisons ($p < 0.05$). Labels correspond to: reverb. medium = reverberation time = 400 ms and reverb. strong = reverberation time = 1,500 ms. Reprinted, with permission, from [3]

$p < 0.01$, $\eta^2 = 0.50$). The plots in Fig. 4.5 depict the average P300 peak amplitude versus reverberation time.

As can be seen in Fig. 4.5, *P300 amplitude* increases with an increase in *reverberation time*. In Table 4.3 the correlations obtained between P300 peaks and all subjective parameters were reported. The spatial distribution of ERP activity will not be analyzed, as it was assumed that a similar pattern—compared to the presented responses in the *Word Experiment* and *Phoneme Experiment*—will also be present in case of sentence long stimuli. As can be seen in Table 4.3, significant negative correlation was attained with *MOS* and the *valence* dimensions. Lastly, a significant main effect with reverberation time was also observed for *reaction time* ($F(2,16) = 11.73$, $p < 0.01$, $\eta^2 = 0.59$).

Table 4.3 Correlation between P300 amplitude and quantitative parameters

Parameter	MOS	A	V	D	L	F
P300	−0.44*	−0.15	−0.40*	0.15	−0.27	0.01

** $p < 0.01$ and * $p < 0.05$
A arousal, *V* valence, *D* dominance, *L* liking, *F* familiarity

4.4 Discussion

In this experiment, the effects of increased reverberation levels on self-assessments of quality by human participants, as well as affective and experience scores were investigated. Inherent human cognitive/neural effects were also observed via EEG *P300 amplitudes* and *reaction times*. As anticipated, subjective quality (*MOS*), experience (e.g., *liking*), and *valence* ratings decreased as *reverberation levels* increased. Interestingly, arousal levels also decreased as reverberation times increased. Given the significant positive correlations observed between arousal and liking, it is conjectured that as reverberation times increased, listening quality decreased and participants became less engaged in the task, and thus were less aroused. In practical conversational situations where reverberation can affect intelligibility, it was expected that increased arousal would be found with increasing RT.

Moreover, participants felt more *dominant* in their judgments of the clean stimuli compared to the stimuli with reverberation. With higher reverberation time, more temporal smearing occurred and resulted in less dominant judgments. As anticipated, participants were more *familiar* with the quality of the clean stimulus, as none of the participants were accustomed to communicating hands-free in an environment with such high reverberation levels. In such cases, the "internal representation/reference" (see Fig. 1.1) of the listener could not account for such distortions. Perhaps if lower reverberation time values had been explored (e.g., between 200 and 500 ms), the listeners would have been more familiar with the introduced distortions.

Regarding the observed cognitive/neural correlates, *P300 peak amplitudes* were seen to be significantly correlated with the MOS and valence parameters, thus showing that the lower the subjective ratings were, the higher the *P300 peak amplitude* and the self-assessed arousal level.

Moreover, increased *P300 amplitudes* were observed as reverberation levels increased, suggesting that participants found the listening task to be less demanding as reverberation levels increased. This was also supported by the decrease in arousal levels as quality decreased. It is believed that an inverse relationship would have been observed if the test had been either an intelligibility or a conversational task, as participants would require greater attentional resources (lower P300 amplitudes) as quality decreased—thus being more in line with practical situations. It can be recommended when performing subjective listening quality tests with reverberant speech that a relevant task should be given to the participants, so that they remain attentive to the spoken content (e.g., what time will the bus arrive?); this is similar to what is done in the context of listening quality assessment for text-to-speech systems. As was previously shown [1, 72, 85, 105], EEG can be used to gather cognitive and Quality of Experience insights for stimulation with sentence-long stimuli in complex listening situations.

Lastly, it was observed that the *response time* versus *reverberation time* curve was non-monotonic. This behavior may have been different, if a lower range of RT had been used, and/or if participants had been given a task to perform while listening to the speech files. As quality decreased to less acceptable values, participants were quicker in judging the listening quality. For intermediate quality levels, judgment took longer, perhaps because participants were hesitant to describe the final quality score.

4.5 Chapter Summary

In this chapter (Chap. 4, *Sentence Experiment*), the generalization of the test set-up was further applied to sentence-long stimuli in complex listening environments (reverberation). The results prove that the combined test set-up is also suitable for stimuli with length that are typically used in standard speech quality perception testing. This constituted the fourth major contribution of this book.

In the next chapter (Chap. 5, *Audiobook Experiment—Constant Quality*), the EEG signal will be analyzed using a different approach, i.e., by analyzing the frequency band power, showing that the quality of presented speech can influence the cognitive state of listeners.

Chapter 5
EEG Frequency Band Power Changes Evoked by Listening to Audiobooks at Different Quality Levels

In the current chapter (Chap. 5, *Audiobook Experiment—Constant Quality*), the frequency band power of the alpha and theta EEG bands will be analyzed. As stimulus material, speech files with the length of audiobooks will be used (for an overview see Fig. 5.1). The degradation factors will be differing bit rate conditions of a speech codec with a constant quality level during each presentation block.

5.1 Introduction

As introduced in Chap. 1, the quality of transmitted media is an aspect of continuous importance to service providers.[1] One reason is the assumption that users would be willing to make more frequent use of media services, if the quality of the transmitted media was higher and that the usage periods were longer [11]. Such assumed or observed differences might coincide with physiological changes in the perceiving user which have not yet been fully understood.

Common methods to determine the quality of media rely on conscious ratings of participant opinion concerning the quality of the presented stimuli. In standard continuous quality scale (CQS) tests, participants are exposed to media of relatively short duration as demonstrated in Chaps. 2 and 3. Longer stimuli may also be presented and judged in a continuous way (cf. the method described in ITU-T Rec. P.880 [106]), whereby a conscious judgment will once again be requested. Whereas such methods provide a reliable and valid means of determining quality, they provide little insight into the physiological processes preceding the quality judgment, which, however, may affect the subjective behavior, e.g., in terms of alertness or media usage duration.

In Chaps. 2 and 3, it was shown that participants sometimes probably do not notice a degraded stimulus, although the brain is in fact processing the degradation on a

[1] This chapter is based on a previous publication; text fragments, tables, and figures are based on Antons et al. [2]. Reprinted, with permission, from [2].

J.-N. Antons, *Neural Correlates of Quality Perception for Complex Speech Signals*, T-Labs Series in Telecommunication Services, DOI 10.1007/978-3-319-15521-0_5

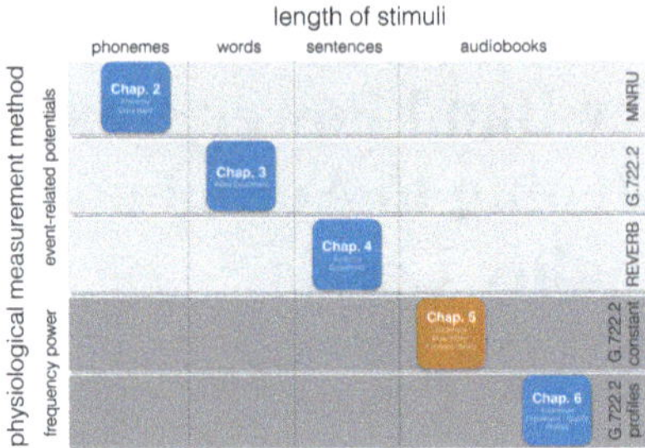

Fig. 5.1 Conducted experiments and structure of this book. Different lengths of stimuli on the x-axis (phonemes, words, sentences, and audiobooks). Physiological measurement techniques on the y-axis (EEG frequency band power and event-related potentials). Applied classes of degradations are color-coded (*grey bars*) and indicated on the *right* (signal-correlated noise introduced by a modulated noise regulation unit (MNRU), bit rate reductions introduced by using different settings of a speech codec in accordance with ITU-T Rec. G.722.2, and reverberation (REVERB) introduced by different room impulse responses. Current chapter is indicated in *orange*. Chapter 5, *Audiobook Experiment—Constant Quality*

non-conscious level. Such neural effects—which become visible in a split-second after stimulus presentation—could be observed for the auditory domain (phonemes and words: [1, 41]) as well as the visual domain (zoom into images: [86], videos: [44]).

In this chapter, it will be examined how these phasic, i.e., short-term changes might eventually lead to tonic, i.e., long-term effects on the user state. In order to address this question, an approach will be investigated to assess the state of the user in relation to fatigue when she/he is being confronted with speech stimuli of different audio bit rate, by analyzing the spectral components of *electroencephalogram* (*EEG*) signals (see Sect. 1.3.1).

In the following analysis, these frequency bands will be assessed for listeners who are confronted with longer (approx. 40 min) speech stimuli of varying quality. Bit rate limitation was utilized as a quality limitation factor, which can also be assessed consciously, e.g., in normal ACR tests. An accompanying ACR test was carried out as a sanity check, in which participants rated the quality of the stimuli without reference.

5.2 Methods

5.2.1 Participants

Eighteen students of the Technische Universität of Berlin participated in the *Audiobook Experiment—Constant Quality* (ten females and eight males; average age = 25.56 years; SD = 3.56; range = 21–31 years old), all of them native German speakers. All participants reported normal auditory acuity and no medical problems. Participants provided their informed consent and received monetary compensation. The experiments were conducted in accordance with ethical principles stated in the Declaration of Helsinki.

5.2.2 Materials

As a test stimulus, a 40-min-long audio recording with information read aloud (by a male speaker) about sights in the city of Berlin, Germany was used.

A recording of recited information was selected as stimulus, as the content had no explicit or implicit emotional content. This kind of auditory stimulation is usually experienced as rather unexciting though stable in terms of emotional effect. This was important, as individual liking or disliking of the content could influence the resultant quality rating.

Two 20-min-long blocks were created, and for each block an undisturbed and a disturbed version were produced. The degradation factor used was a limitation of the bit rate of the codec described by ITU-T Recommendation G.722.2 [10]. The difference between the high quality (WB; wideband) and lower quality (G7; codec ITU-T Recommendation G.722.2 with 6.6 kbit/s) was the bit rate. Each participant listened to the whole 40 min, with one block degraded and the other part non-degraded (see Fig. 5.2 for an exemplary course of quality for one participant).

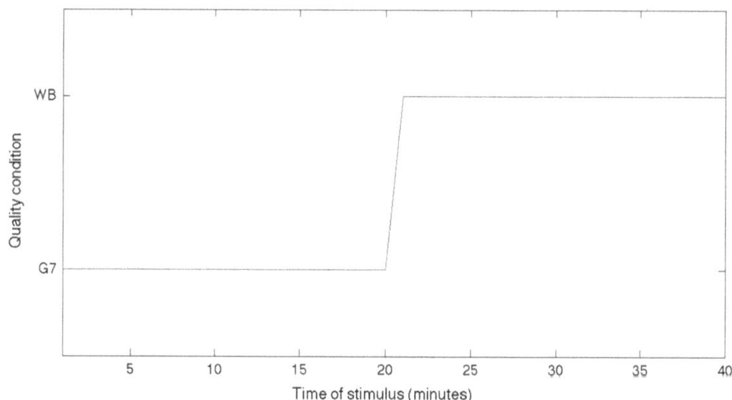

Fig. 5.2 Exemplary course of quality for one participant. In total a 40 min long stimulus was used, which were presented in two blocks in different quality levels (WB = high quality, G7 = low quality). The order of blocks (WB–G7 vs. G7–WB) was randomized between participants. Reprinted, with permission, from [2]

The order of blocks was randomized so that each participant started with either the high-quality block or the block with lower quality. The following combinations were possible; (1) first high quality then low quality (WB–G7) and (2) first low quality then high quality (G7–WB).

5.2.3 Experimental Design

In an introductory phase, it was explained to the participants what their task was and what would happen during the *Audiobook Experiment—Constant Quality*.

The task was to rate the perceived quality of the stimulus on a ACR scale ranging from excellent (9) to bad (1) (see Sect. 1.2 for the detailed explanation of subjective rating methods).

Their task was to rate the perceived quality of the stimulus. After this, a questionnaire was handed out to the participants. Using this questionnaire, demographic data, information about the physical condition of the participants, and the informed consent of the participants were obtained. As there were no irregularities regarding the physical condition and mental state of the participants, this aspect will not be further addressed. While the participant were filling in the questionnaire, the preparation of the electrodes took place.

During the main test, participants had to listen to the presented information. They were not told in what order of quality the parts would be presented. After 9, 17, 25, and 33 min, participants had to rate the perceived quality of the stimulus on an eleven-grade numerical quality scale as proposed by the ITU-T Recommendation P.910 [18]. The scale appeared on a screen placed in front of the participants and they had to indicate the quality on the screen by mouse click (from excellent (9) to bad (1)). Between the ratings, the screen was blank. After the main test, participants had to answer four questions regarding the content of the information that had been read aloud. They were promised some extra monetary compensation, if these questions were answered correctly. This was intended to encourage the participants to listen more carefully. As the questions were just intended as a cover story, two easy and two difficult questions were asked.

5.2.4 Electrophysiological Recordings

The EEG (Ag/AgCl electrodes, Brain Products GmbH, Garching, Germany) was recorded continuously from 7 standard scalp locations according to the 10–20 *system* (FZ; Cz; Pz, P3-4 and O1-2) [65]. The reference electrode was placed on the tip of the nose. Impedances were kept below 10 kOhm. On-line filtering was not carried out. The signal was digitized with a 16-bit resolution and a sampling rate of 1,000 Hz.

5.2.5 *Statistical Analysis*

Only effects significant at the alpha level $p < 0.05$ that were relevant to the hypotheses are reported in the following analysis.

5.2.5.1 Quality Ratings

Ratings of the degraded part and ratings of the non-degraded part were compared by means of a dependent t-test.

5.2.5.2 Frequency Band Power Analysis

Off-line signal processing of the EEG data was carried out using the MATLAB® toolbox EEGLAB [92]. The raw EEG data was low-pass filtered with a finite impulse response filter (filter with the critical frequency of 40 Hz). For each frequency band, the electrode with the highest band power was automatically selected participant-wise and used for the further analysis. The power of all frequency bands for each participant and the two time intervals per block (disturbed and undisturbed quality) were calculated, one for the first 10 min of the block (initial 10 min) and one for the second half (final 10 min). In total, four power values per frequency band and participant were registered.

An analysis of variance (ANOVA) for repeated measures with *degradation intensity* and *time of measurement* as the independent variables, as well the power of the the *theta band* and the *alpha band* as the dependent variables—was carried out.

In order to check whether the power values for alpha and theta bands were already different from one another during the first presented block, another ANOVA between participants with adjusted alpha levels was calculated. As independent variables, the power of the *alpha* and *theta* frequency band, and as dependent variables, the *type of stimulus during the first block* were used.

The main focus of analysis in this experiment was placed on the practical application of physiological parameter, thus, purely the power of the theta and alpha frequency band were analyses at the electrode with the highest power values of the respective frequencies. An analysis and plot of the spatial distribution of frequency power values at several electrodes was not performed, which remains as a task for further research.

5.3 Results

5.3.1 Quality Ratings

The analysis of the subjective rating during the main test revealed a significantly different rating depending on the *quality* of the presented stimulus (Fig. 5.3). As anticipated, the quality of the undisturbed stimulus was rated significantly better ($T = 5.92$, $p < .01$).

5.3.2 Frequency Band Power Analysis

Within the EEG data, two significant effects were found. For the *degraded quality*, an increase in the *theta* band was found ($F(1,17) = 9.85$, $p < 0.05$, $\eta^2 = 0.36$). Participants had significantly increased *theta* band activation while listening to the disturbed part. As can be seen in Fig. 5.4, there was no effect for early vs. late parts (start and end of each block) ($F(1,17) = 1.35$, $p = 0.26$, $\eta^2 = 0.07$).

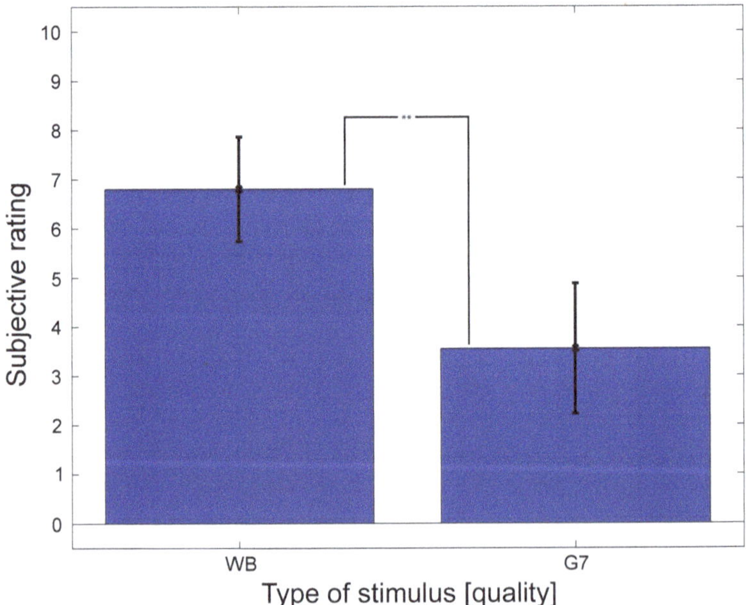

Fig. 5.3 Grand average of subjective ratings. Undisturbed (WB) and disturbed stimulus (G7). Higher values indicate higher quality. Whiskers denote standard deviation. *Asterisks* denote the significance level of $p < 0.01$. Reprinted, with permission, from [2]

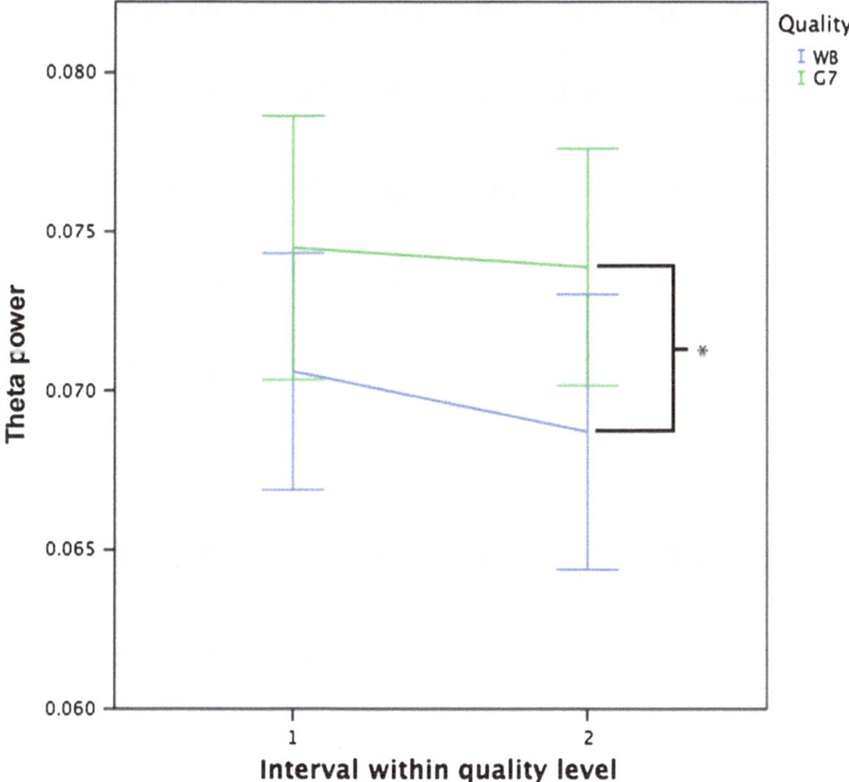

Fig. 5.4 Grand average theta frequency activation. Undisturbed (WB) and disturbed stimulus (G7) for the initial 10 (number 1) and final 10 min (number 2) of each corresponding block. Each participant listened to a 40 min-long stimulus, which was presented in two blocks with different quality levels (WB, G7). The order of blocks (WB–G7 vs. G7–WB) was randomized between participants. For this grand average, the values for the blocks of different quality were averaged—independently from the initial position for each individual participant. Whiskers denote standard errors. *Asterisk* denotes the significance level of the main effect introduced by the quality factor on the theta frequency band power ($p < 0.05$). Reprinted, with permission, from [2]

For the *two time intervals* of analysis (start vs. end of the block), a significant increase in *alpha* band activity was found ($F(1,17) = 5.66$, $p < 0.05$, $\eta^2 = 0.25$). For both conditions (disturbed and undisturbed signal), activity in the *alpha* band increased during presentation; participants had a higher power in the *alpha* band at the end (WB = 0.0662 and G7 = 0.0669) of the stimulus compared to the beginning (WB = 0.0616 and G7 = 0.0640). Alpha power tends to be higher for the disturbed version from beginning on, as depicted in Fig. 5.5.

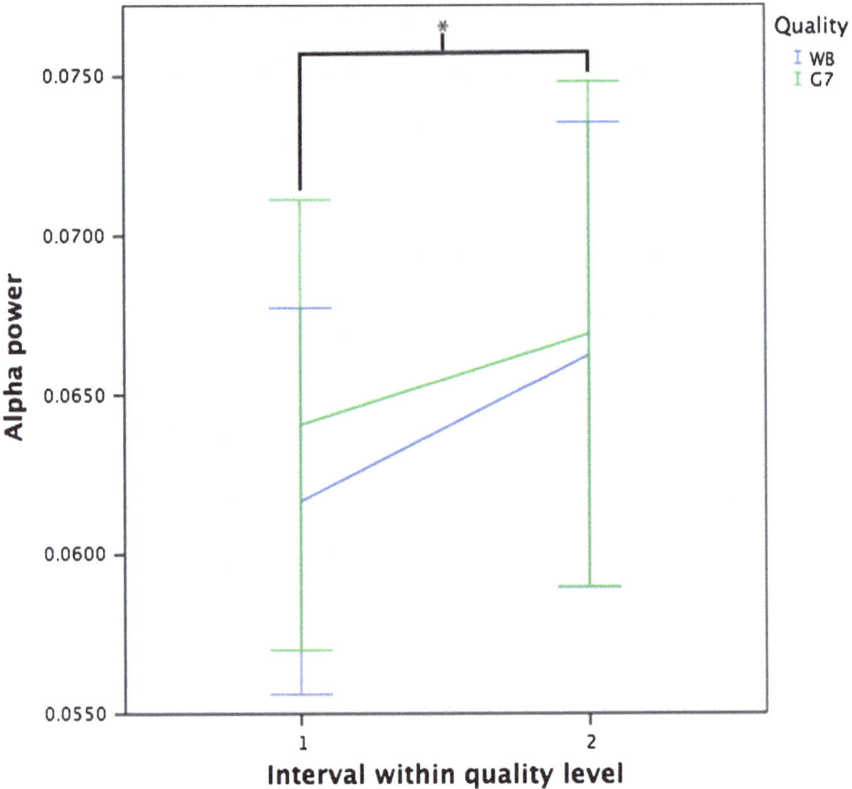

Fig. 5.5 Grand average alpha frequency band power. Undisturbed (WB) and disturbed stimulus (G7) for the initial 10 (interval 1) and final 10 min (interval 2) of each corresponding quality level. Whiskers denote standard errors. Each participant listened to a 40-min-long stimulus, which was presented in two blocks with different quality levels (WB, G7). The order of blocks (WB–G7 vs. G7–WB) was randomized between participants. For this grand average, the values for the blocks of different quality were averaged independently from the initial position for each individual participant. *Asterisk* denotes the significance level of the main effect introduced by the factor: time interval of measurement on the alpha frequency band power ($p < 0.05$). Reprinted, with permission, from [2]

At the end of the presentation, *alpha* power for the undisturbed stimulus almost converges to the level of the disturbed value, thus leading to a non-significant overall effect of stimulus quality ($F(1,17) = 0.34$, $p = 0.56$, $\eta^2 = 0.02$). In addition, there is also no significant interaction between quality and time interval (start vs. end) ($F(1,17) = 1.84$, $p = 0.19$, $\eta^2 = 0.09$).

The analysis, whether the alpha and theta power already differ between participants during the first block, was non-significant (alpha: $F(1,17) = 0.02$, ns, and theta $F(1,17) = 0.64$, ns).

5.4 Discussion

As expected for the *subjective rating* of listeners, it was found that the quality judgment of the degraded stimulus was rated lower in comparison to the non-degraded stimulus.

This part of the experiment was considered to be a check, i.e., whether the quality variation was perceptible and showed that the degradation quality of the assessed stimuli was well above the threshold of a conscious judgment of the perceived degradation.

Based on the analysis of *EEG frequencies*, quality was observed to have an effect on the power of the *theta* frequency band. Due to the fact that an increases in this band are known to reflect drowsiness, it has been concluded that such an increase while listening to the disturbed stimulus is in fact an increase of drowsiness due to impaired information processing. Apparently, test participants listening to the degraded speech stimuli became more fatigued.

In respect to the *alpha* frequency band, a significant increase in spectral power was observed as well. Higher power in the alpha frequency band for the second time interval of the analysis could correspond to a decrease in alertness, probably due to the time spent on the task. The longer the participants listened to a stimulus, the more tired they became. This general tendency was observed both for the degraded and for the non-degraded speech stimuli. However, the tendency was evident that participants listening to the disturbed stimulus had already entered a decreased alertness state during the first 10 min, which was not the case when listening to non-degraded stimuli.

The non-significant comparison between participants of the two quality levels within the first presented block seem to be attributable to the high variance between participants. Therefore, a repeated measurement design (within participant design) is appropriate, as it accounts for these differences by analyzing variances within the individual participant's data.

Surprisingly, the changes in alpha and theta band power were specific to two different aspects, i.e., quality level and time-on-task, although they indicate a similar physiological state.

As the theta band is known to increase during high workload conditions (tasks with a high mental workload), it could be the case that the increase during the presentation of the low-quality segments caused a momentarily high workload, and therefore, also an increase in this band [47]. At the same time, the alpha band is known to be suppressed during high workload conditions and elevated in the long run during the onset of fatigue and sleep.

5.5 Chapter Summary

In this chapter (Chap. 5, *Audiobook Experiment—Constant Quality*), a new method of analyzing the EEG signal was implemented. The frequency band power was analyzed, indicating that the quality of the presented speech can probably influence the cognitive state of listeners. The results showed that long speech stimuli presented in low quality will lead to a reduced cognitive state of listeners. This was done for 20-min-long blocks of an audiobook in which the quality was constant within one block.

In the next chapter (Chap. 6, *Audiobook Experiment—Quality Profiles*), this approach will be applied to audiobook long speech stimuli with different quality profiles. Accordingly, the quality within one block will be altered.

Chapter 6
EEG Frequency Band Power Changes Evoked by Listening to Audiobooks with Varying Quality Profiles

In the previous chapter (Chap. 5), the frequency band power of the alpha and theta EEG bands was analyzed. Audiobooks were used as stimulus material. The *Audiobook Experiment—Constant Quality* showed that lower quality of long speech stimuli can probably influence the cognitive state of listeners.

In the current chapter (Chap. 6, *Audiobook Experiment—Quality Profiles*), the method applied here will be transferred to audiobooks with different quality profiles (for an overview see Fig. 6.1). Within the presented speech stimuli, the quality will be altered by bit rate changes of the used speech codec.

6.1 Introduction

In the previous chapter (*Audiobook Experiment—Constant Quality*), the question was answered as to whether these phasic, i.e., short-term changes eventually might lead to tonic, i.e., long-term effects on the user state.[1] It could be demonstrated that the reduced bit rate of codecs resulted in a reduced vigilance state for listeners, and therefore, in impaired information processing.

As was discussed in the previous experiment, the elevated power in the theta frequency band can be associated with fatigue and with high workload situations. As this experiment intends to measure a reduced cognitive state (drowsiness), no theta band frequency band power will be analyzed, thereby minimizing the risk of confounding the influence of the two concepts.

In the following, the *alpha frequency band power* of listeners who are confronted with longer (approximately 40 min) speech stimuli of varying quality will be assessed. As a quality variation, a bit rate limitation which influences quality ratings was used. As this variation can also be estimated by instrumental methods the E-Model and POLQA were used. Therefore, the estimated quality was calculated as a value which

[1] This chapter is based on a previous publication; text fragments, tables, and figures are based on Antons et al. 2013b [4]. Reprinted, with permission, from [4].

J.-N. Antons, *Neural Correlates of Quality Perception for Complex Speech Signals*, T-Labs Series in Telecommunication Services, DOI 10.1007/978-3-319-15521-0_6

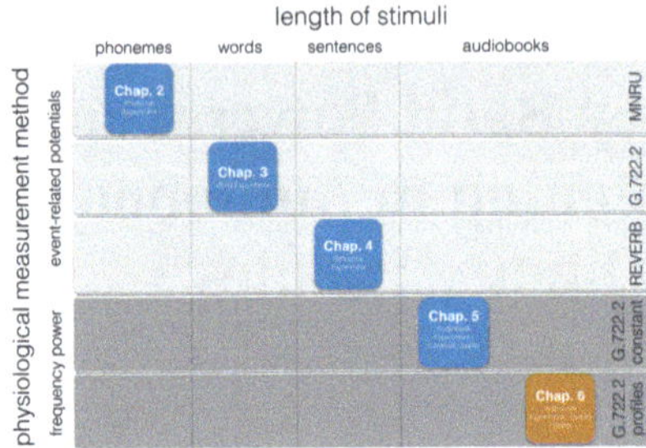

Fig. 6.1 Conducted experiments and structure of this book. Different lengths of stimuli on the x-axis (phonemes, words, sentences, and audiobooks). Physiological measurement techniques on the y-axis (EEG frequency band power and event-related potentials). Applied classes of degradations are color-coded (*grey bars*) and indicated on the *right* (signal-correlated noise introduced by a modulated noise regulation unit (MNRU), bit rate reductions introduced by using different settings of a speech codec in accordance with ITU-T Rec. G.722.2, and reverberation (REVERB) introduced by different room impulse responses. Current chapter is indicated in *orange*. Chapter 6 *Audiobook Experiment—Quality Profiles*

could be compared with the values of the frequency band power collected here. It is known that quality can be only be estimated; for a real sanity check, a subjective quality test would have to be performed. Nevertheless, it could be shown that the estimated quality is similar to the measured quality values in a subjective rating experiment [25].

6.2 Methods

6.2.1 Participants

Twelve students participated in the experiment (3 females and 9 males; average age = 26.33 years; SD = 3.62; range = 20–34 years old), all of them native German speakers. All participants reported normal auditory acuity and no medical problems. Participants provided their informed consent and received monetary compensation.

6.2.2 Stimuli

As a test stimulus, a 40-min-long audio recording with information read aloud (by a male speaker) about sights in the city of Berlin, Germany was used. Two 20-min-long blocks were created, and for each block, two quality profiles using varying bit rates of the codec described by the ITU-T Recommendation G.722.2 [7] were produced. First, a constant profile with the bit rate 12.65 kbit/s for the complete block; second,

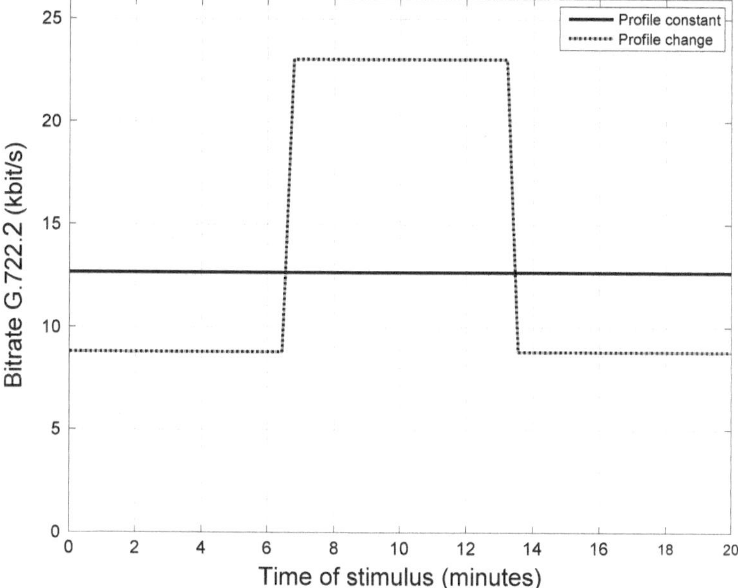

Fig. 6.2 Profiles used during the experiment. Profile constant with one constant bit rate and profile alteration with a changing bit rate within the block. All participants listened to both profiles, which resulted into a total listening duration of 40 min. The order of blocks (cha-con vs. con-cha) was randomized across participants. Reprinted, with permission, from [4]

a varying profile changing the bit rate after 6 min and 40 s from 8.85 to 23.05 kbit/s; and then after 13 min and 20 s back to 8.85 kbit/s (see Fig. 6.2).

Both profiles resulted in an approximately similar number of transmitted kbit for the complete block, 16,300 transmitted kbit for the changing profile, and 15,180 kbit for the constant profile.

The transmitted kbit per experimental block were selected to be similar, and the intention was to analyze the impact of a stimuli with similar requirements for transmission on a subjective scale as well as the corresponding physiological responses (variations in frequency band power).

Each participant listened to the whole 40 min, with one block constant and one with changing bit rates. The order of blocks was randomized.

6.2.3 *Experimental Design*

Participants had to complete questionnaires requesting demographic data and information about physical conditions. As no abnormalities were reported, detailed information about these aspects will not be provided here. While participants were filling in the forms, the electrodes were prepared.

During the test, participants had to listen to the prepared audio clips. The order of the two profiles which were presented was not told to the participants. In order to draw the attention of the participants even more on the audio clip, four content-related questions were asked at the end, and correct answers were rewarded with an additional compensation. In this context, two easy and two more difficult questions were asked, however, this aspect will not be covered in this book.

6.2.4 Electrophysiological Recordings

The EEG (actiCap electrodes, Brain Products GmbH, Garching, Germany) was recorded continuously from 32 standard scalp locations according to the 10–20 *system* (FP1-2; Fz, 3-4, 7-8; FC1-4; T7-8; Cz, 3-4; CPz, 1-2, 5-6; Pz, 3-4, 7-8; POz, 9-10; Oz, 1-2) [65]. Impedances were kept below 10 kOhm. On-line filtering was not carried out. The signal was digitized with a sampling rate of 1,000 Hz.

6.2.5 Statistical Analysis

6.2.5.1 Instrumental Quality Estimation

In order to be able to compare the physiological parameters registered in the experiment, instrumental quality estimations using the E-Model and POLQA were calculated. Of course, it is not possible to estimate judgment by instrumental methods, so these estimations can merely count as sanity checks, but as shown in, e.g., [25] and [30], the estimated quality can be considered close to the actual ratings of users. As one of the goals of this experiment was to determine whether the variation of EEG frequency band power is indicative of stimulus parameter, which would also influence the perceived quality, instrumental quality estimation was used. The estimated quality was calculated using the first 8 s of each bit rate level for each instrumental quality estimation method (E-Model and POLQA).

E-Model

The *E-Model* is a parameter-based instrumental model first published in 1997 [107]. The output of the model is an estimation of transmission quality, named the R-value, ranging from $R = 0$ to $R = 100$. The latter range is for narrowband and can also be extended to wideband [25] and super-wideband [108], resulting in an extended E-Model rating scale up to 129 and 179, respectively. A major feature of this model is the use of transmission impairment factors that reflect the effects of modern signal processing devices [34]. The mean opinion score (MOS) of the quality estimation can be calculated using the R-value.

POLQA

POLQA (*perceptual objective listening quality assessment*) is a signal-based instrumental model [109]. The goal of the program was to select an intrusive speech quality model suitable for NB up to super-wideband (S-WB) connections (50–14,000 Hz) and to compensate the defects observed in the PESQ model [110] and [111]. The measurement algorithm is a full reference model, which operates by conducting a comparison between a known reference signal and a captured degraded signal that was sent through a transmission system. The output of POLQA is an estimated quality (MOS). To compute the estimated quality of stimuli with the POLQA model, only two sentences of the whole stimulus per bit rate level were used. Thereby, it was made sure that the signals met the requirements and that the quality of the transmission system could be estimated properly. In contrast to POLQA—which is suitable for up to super-wideband signals, extensions for the PESQ model to wideband are available [112].

6.2.5.2 Frequency Band Power Analysis

Off-line signal processing of the EEG data was carried out using the MATLAB® toolbox EEGLAB [92]. The raw EEG data were bandpass-filtered with a finite impulse response filter (filter with the critical frequencies of 1 and 44 Hz). The analysis was conducted using the data of electrode CPz. For the analysis of alpha frequency band power, the power in the band 8–13 Hz was computed. The power of the alpha frequency band for each participant, the two quality profiles (constant and change) and for three time intervals per profile (time 1: minute 3–9, time 2: minute 9–15 and time 3: minute 16–20) was calculated. The first interval (minute 1–2) served as a baseline and the following three intervals were of equal length (6 min each). In total, six power values per participant that were baseline corrected, meaning that the alpha band power of a 2-min long baseline from the beginning of the respective profile was subtracted from each value, were registered. The results were analyzed by calculating a repeated analysis of variance (ANOVA), with *profile type* and *time interval within profile* as the independent variables, and the *alpha band power* as the dependent variable.

Due to the fact that the analyses of power values for *alpha* and *theta* bands were not significantly different between participants within the first block presented in the previous chapter (*Audiobook Experiment—Constant Quality*), this analysis was not performed here.

It has to be mentioned again, that focus in this experiment—as in Chap. 5, *Audiobook Experiment—Constant Quality*—was placed on the practical application of physiological parameter, thus, purely the power of the alpha frequency band were analyzed at the electrode with the highest power values. An analysis and plot of the spatial distribution of frequency power values at several electrodes was not performed, which remains as a task for further research.

6.3 Results

6.3.1 Instrumental Quality Estimation

The results of the instrumental quality estimation have been presented in Table 6.1.

The estimated MOS rises with the bit rate. Also, the ratings of the *WB-E-Model* and *POLQA* are similar to each other, except for the 8.85 kbit/s condition, for which the WB-E-Model reaches a lower MOS value. The average values per profile resulted in an almost similar average quality estimation for POLQA (average value profile constant = 4.09 and profile change = 4.01). For the WB-E-Model, the averaged estimated quality values are higher for the constant profile (profile constant = 4.02 and profile change = 3.63).

6.3.2 Frequency Band Power Analysis

Within the EEG data, a non-significant tendency of increased *alpha frequency power* for the profile type *profile type* ($F(1, 12) = 3.38$, $p = 0.09$) was found.

As can be seen in Fig. 6.3, the mean band power for the profile *change* was higher (average value = 2.08) compared to the profile *constant* (average value = 1.53) for all time intervals.

In addition to this, a significant effect of the *time interval within the profile* was found ($F(2, 24) = 3.64$, $p = 0.04$). For the profile constant, a steadily increasing alpha value was observed (interval 1 = 1.4, interval 2 = 1.55, and interval 3 = 1.65). By contrast, it was observed that—for the second time point of profile change—the alpha values decreased and then increased again (Fig. 6.3, green line, interval 1 = 1.97, interval 2 = 1.88, and interval 3 = 2.41). The pairwise post-hoc comparison revealed a significant difference between the first and the third intervals averaged over both profiles ($p < 0.05$).

No interaction of profile type and time interval within the profile was observed ($F(2, 24) = 0.76$, ns).

Table 6.1 Results of the instrumental quality estimation using the WB-E-Model and POLQA

Codec	MOS rating	
	WB-E-Model	POLQA
G.722.2 @ 8.85 kbit/s	3.29	3.82
G.722.2 @ 12.65 kbit/s	4.02	4.09
G.722.2 @ 23.05 kbit/s	4.32	4.4

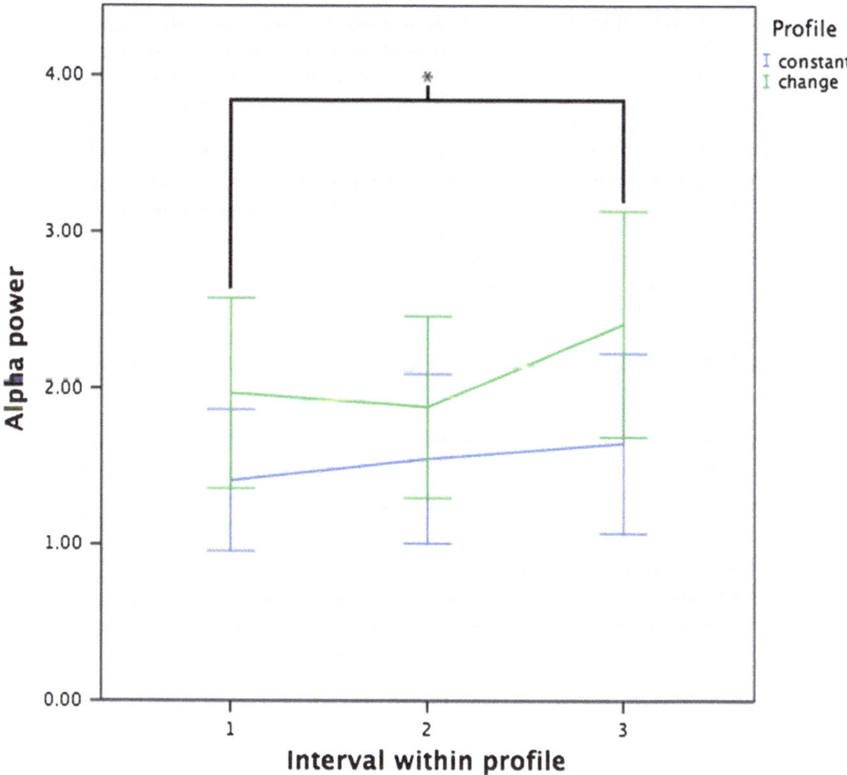

Fig. 6.3 Grand average of alpha frequency band power for electrode CPz. Profile constant with one constant bit rate and profile change with changing bit rate within the block, plotted for the three intervals per profile. Whiskers denote standard errors. *Asterisks* denotes the significance level of the pairwise post-hoc comparison ($p < 0.05$). Reprinted, with permission, from [2]

6.4 Discussion

The instrumental quality estimation resulted in approximately similar values using *POLQA* and the *WB-E-Model*, except for the lower bit rate condition (8.85 kbit/s) for which the value generated by the *WB-E-Model* was lower. As anticipated, the estimated quality rose with the bit rate. This analysis was considered to be a check if the intended quality variation effect the estimated quality judgements.

For the analysis of *alpha frequency band power*, a significant effect of the *time interval within the profile* was found. The increase over time is considered a time-on-task effect. As an increase of this band is known to reflect drowsiness, the increase of band power in the experiment presented here is considered to correspond to an increase of drowsiness as well. Therefore, the longer participants listened to the stimuli, the more fatigued they became, which also lowered their vigilance level. The non-significant tendency of increased alpha frequency power for the *profile type* is evaluated as a state of drowsiness due to the two low quality intervals. For profile change higher vigilance level were shown, this could be due to the fact that the intervals with lower bit rates had a greater impact on the cognitive state than the ones with higher bit rates. This effect is already known from the analysis of subjective data. Following [113], quality profiles with a very low quality segment were judged being from lower quality compared to profiles with constant medium quality.

Participants could recover during the intervals with higher bit rates but were less vigilant for the overall profile change (Fig. 6.3, dashed line).

6.5 Chapter Summary

In this chapter (Chap. 6, *Audiobook Experiment—Quality Profiles*), EEG frequency band power was analyzed, demonstrating that the quality of presented speech can influence the cognitive state of listeners. In combination with the results presented in Chap. 5, *Audiobook Experiment—Constant Quality* this constituted the fifth contribution of this book. The results showed that a period of high bit rate audio inserted into a low bit rate stimulus can probably increase the vigilance of listeners on a time scale of minutes. However, a strong time-on-task effect was present and led to increased alpha frequency band power at the end of the listening task.

Chapter 7
General Discussion and Future Work

The general discussion will be structured similar to this book (for an overview see Fig. 7.1) and the presented contributions:

- Implementation of a test set-up combining neurophysiological and subjective quality assessment methods for speech quality perception testing (Chaps. 2, 3, 4, 5, and 6).
- The proof that this test set-up functions with short speech stimuli (phonemes) and a generic quality impairment, i.e., signal-correlated noise (Chap. 2).
- A successful transfer of this test method to longer speech stimuli (words) with a more realistic quality impairment, i.e., reduced bit rate of a speech codec (Chap. 3).
- The proof that this technique works with stimuli of standard length for speech quality assessment (sentences) and an environmental-dependent quality impairment, i.e., reverberation (Chap. 4).
- Investigation of the impact of a speech compression algorithm with reduced bit rate on the cognitive state of listeners for speech stimuli of long duration (audiobooks) in constant (Chap. 5) and varying quality conditions (Chap. 6).

7.1 General Discussion

This book investigated the usefulness of a combined test set-up using physiological parameter and subjective quality ratings for analyzing human speech quality perception.[1] The objective was to investigate whether ERP components and EEG frequency band powers, specific to the detection of degradations could be identified, potentially

[1] Parts of this chapter have been previously published; text fragments are based on Antons et al. [1], reprinted, with permission, from [1], Antons et al. [2], Reprinted, with permission, from [2], Antons et al. [3], reprinted, with permission, from [3], Antons et al. [4], reprinted, with permission, from [4], and Antons et al. [5], with kind permission from Springer Science+Business Media: [5].

J.-N. Antons, *Neural Correlates of Quality Perception for Complex Speech Signals*, T-Labs Series in Telecommunication Services, DOI 10.1007/978-3-319-15521-0_7

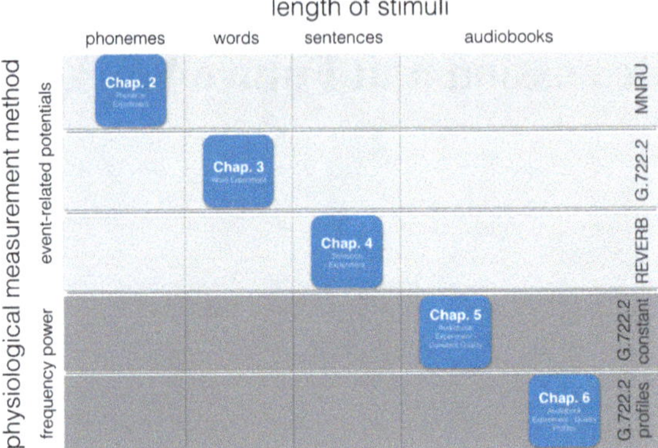

Fig. 7.1 Conducted experiments and structure of this book. Different lengths of stimuli on the x-axis (phonemes, words, sentences, and audiobooks). Physiological measurement techniques on the y-axis (EEG frequency band power and event-related potentials). Applied classes of degradations are color-coded (*grey bars*) and indicated on the *right* (signal-correlated noise introduced by a modulated noise regulation unit (MNRU), bit rate reductions introduced by using different settings of a speech codec in accordance with ITU-T Rec. G.722.2, and reverberation (REVERB) introduced by different room impulse responses

also for non-conscious processing steps when listening to degraded speech files, and to test the applicability of the method in a realistic application scenario. Fife experiments were carried out to perform the analysis.

7.1.1 Phoneme Experiment

In the *Phoneme Experiment*, it was shown that the participant *detection rate* in the oddball paradigm experiment reached the 50 % threshold at the same SNR level at which test participants also rated the quality significantly worse in the *opinion test*. The *reaction time* in the oddball paradigm was significantly higher for the high SNR condition, reflecting the cognitive effort required to process the (subtle) degradation. The *P300 peak latency* showed that the lower the degradation level, the later a P300 was evoked, which is most likely due to the same reason, namely, the higher cognitive effort involved in detecting the degradation. In turn, the stronger the degradation, the higher the *P300 peak amplitude*. Using *LDA classifiers* on the EEG signals, it could be shown that patterns of brain activation which were similar to those for detected degradations could also be observed in trials where the participants did not report a degradation. It is likely that small degradations are non-consciously processed in a way similar to larger ones, although they do not result in the same conscious rating. The brain accomplishes the given experimental task predominantly in the way as expected. Strongly degraded stimuli were assessed more quickly compared to the weaker degradations. The stronger the degradation, the earlier the maximum amplitude of P300, and the higher its amplitude. The expected scalp potential distribution and peak latency match findings in literature and other studies [80]. One unusual feature is the detection of degradation-specific brain responses to weak degradations which were not reported at the behavioral level. Thus, ERP analysis might provide objective evidence for the non-conscious engagement of brain processes by minor stimulus degradations, which could eventually influence the listeners appreciation of stimulus quality during long-term confrontation with such degraded material.

The subjective rating behavior was further analyzed in the *Length Influence Experiment*. The results substantiated that stimulus length had a significant impact on subjective responses, with stimuli of word or sentence length being rated significantly worse than phoneme stimuli. This implies that stimuli of at least word length should be used in subsequent subjective experiments in order to reflect realistic usage scenarios of subjective test methods. The type of headphone did not have a significant effect. Therefore, in-ear headphones, which are more common in EEG studies, can be used.

7.1.2 Word Experiment

In the *Word Experiment*, similar results could found, this time in respect to coding distortions rather than signal-correlated noise. It should be noted, however, that the subtlety of the degradation did not affect the latency of the P300 for coding distortions in words. This leads to the question as concerning which stimulus modalities and types of degradations in ERP-based analysis can be applied. Considering the initial investigations reported in [42–44, 85, 114], it can be postulated that the method is equally applicable to auditory, visual, and audiovisual stimuli. With respect to the latter, further insight into the quality integration process of auditory and visual perception can be anticipated, and may be obtained by the method presented here.

7.1.3 Sentence Experiment

The *Sentence Experiment* has explored cognitive, affective, and experiential factors inherent to participants when asked to perform listening tasks in the context of speech quality assessment; these insights had been previously non-existent in the literature on quality assessment. Focus was placed on the Quality of Experience (QoE) assessment of reverberant speech, thus simulating hands-free communication. Increased *P300 amplitudes* were observed as reverberation levels increased, suggesting that participants found the listening task to be less demanding as reverberation levels increased. This was also supported by the decrease in arousal levels as quality decreased. It is expected that the use of physiological parameter will lead to improved room acoustic characterization algorithms and more effective listening tests. This could be accomplished by using physiological parameters during such tests, making use of parameters that are probably more sensitive and are able to measure presumably non-conscious brain responses and possible long-term influences on the cognitive state of participants.

7.1.4 Audiobooks Experiment: Constant Quality

The *Audiobook Experiment—Constant Quality* showed that listening to long auditory stimuli contaminated by a degradation induced a brain state similar to drowsiness, which is known to lead to impaired information processing [48]. Nonetheless, this experiment was limited to speech stimuli with one particular type of degradation, whereby the level of degradation was well above the threshold beyond which test participants would consciously rate a stimulus as being degraded. It is left to future research to determine whether comparable physiological responses can also be observed for stimuli whose level of degradation—which can be observed with direct rating procedures such as ACR—is small and perhaps non-significant. In respect to EEG analysis, measurement of subtle differences will be facilitated by including the power of the alpha frequency band in the analysis.

7.1.5 Audiobooks Experiment: Quality Profiles

The *Audiobook Experiment—Quality Profiles* showed that listening to long auditory stimuli with varying quality level introduced a brain state similar to drowsiness. First, it was shown that a varying and a constant bit rate introduced a state of drowsiness. Furthermore, it was observed that low quality intervals had a stronger impact on the cognitive state, which resulted in a lower vigilance state. By means of continuous physiological measurement (alpha frequency band power analysis), it was possible to observe the potential effect that a period of high bit rate audio inserted into a low bit rate stimulus increased the vigilance of listeners on a time scale of minutes. This effect would probably have remained undetected using merely standard techniques of subjective testing. These results can make a significant contribution to better understanding how the quality of media influences the cognitive state of costumers. Whenever the vigilance state of listeners plays an important role, continuous measurement techniques such as the electroencephalographic analysis of alpha frequency band power can promote additional information about, e.g., variations of cognitive states on a time scale of minutes.

7.1.6 Considerations on the Quality Formation Process

In this paragraph, the aforementioned results will be set in relation to the presented quality formation process (Fig. 7.2).

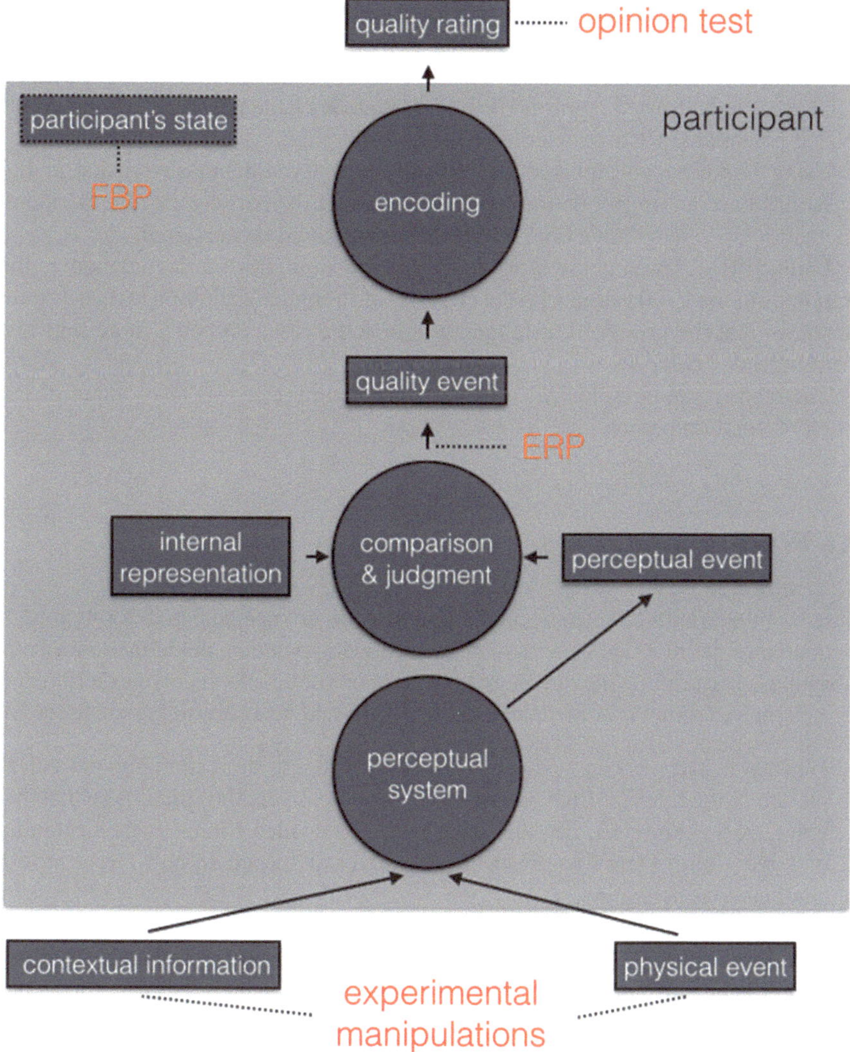

Fig. 7.2 Simplified version of the quality formation process (based on [12, 14, 15, 32]). *Circles* represent perceptual processes, *boxes* represent storage for different types of representation. Note that boxes outside of the participants represent input and output including the incoming signal. The comparison of internal representation and perceived event will effect a quality rating after encoding. Marked in *red* are the parameters used in this book to influence the input (experimental manipulations) and the measuring criteria (event-related potentials/ERP), alpha frequency band power (FBP) and subjective ratings (opinion test). The (cognitive) state of the participant can influence all stages of the formation process. Note that this picture does not include either the detailed anticipatory process or the detailed comparison and judgment process (see [15] for a more detailed model). For a detailed description see text in Sect. 1.1

Experimental manipulation influenced the inputs of the process: the physical event (stimulus) and the contextual information (test environment and task).

Within the process, there were two measurement possibilities which corresponded to the two methods used in this book: alpha frequency band power (FBP) and event-related potential (ERP).

Using the measure of alpha frequency band power, it was demonstrated that quality variations have an impact on the (cognitive) state of the participants. Furthermore, fluctuation of state within a time frame of minutes could be measured.

Using ERPs, it was possible to determine the strength of a degradation as the amplitude of the P300 varied with the intensity of strength. In addition to this, it could be shown that the presented technique was in some cases probably more sensitive than subjective behavior.

The quality rating collected in form of an opinion test was also output of the quality formation process.

7.2 Practical Guidelines

Based on the presented results and the practical experience gathered during implementation of the reported experiments, the following practical guidelines have been developed. These guidelines are intended to support the development of experimental set-ups for experiments at the intersection of QoE and neurophysiological research.

1. When you start working with a new stimulus type, use short stimuli to get a clear picture of the relevant ERP variations. If possible, select the stimulus so that that you have a clear onset, i.e., the onset of audio or video stimuli at the beginning of a recording. Please be aware that audiovisual speech stimuli rarely have a simultaneous beginning.
2. Start with a minimal stimulus set: (i) concerning stimulus length, perform tests with shorter stimuli before you aim for longer ones, (ii) use only a reduced set of speakers and sentences for auditory experiments and a reduced set of scenes for (audio)visual experiments, respectively, and (iii) use preferably one class of degradation.
3. Select only a few levels of degradation (e.g., three noise levels) instead of the full spectrum in order to reduce time expenditure. Correspondingly, determine individual levels of degradation intensity for each participant. It is best to aim for a similar percentage of detected versus non-detected levels of degradation for each participant.
4. Control the experimental environment closely. If existent use, e.g., ITU recommendations (such as [19]), and if appropriate, reduce the suggested set-up (see guideline 1 and 3).

5. Use one of the established set-ups for presentation, i.e., the oddball paradigm with short stimuli [69].
6. Adhere to established analytical paradigms at the beginning—data on brain activity tends to be overwhelming and polysemous, as it represents a variety of influences in addition to the ones you are concerned with in with your study. Established approaches developed over the years attempt to rule out as many of these variations as possible.

7.3 Future Work

One point of future work will be to analyze the impact that the presumably non-conscious detection of degradations may have on perceived quality aspects using other physiological methods such as *heart rate variability* (*HRV*), *functional magnetic resonance imaging* (*fMRI*), *near-infrared spectroscopy* (*NIRS*), and non-physiological indices (self-reported load indices). In the long run, it can be anticipated that ERP-based analysis will be one of several methods providing insight into the quality perception and judgment processes which are still not well understood. In the future, a new compression or transmission method might be declared "subjectively lossless" only if the brain activity of the participant shows no difference compared to activity during perception of the original signal.

In an additional analysis—a single-trial analysis—of the EEG data it could be shown that the attenuated P300 was partly due to the increasing jitter of the P300 component across participants [115] (see [83] for an overview on attenuated ERP components due to jittering effects). For the approach of this book, the grand average of physiological and subjective parameters was used and allowed a valid comparison at the group level (as is done in most testing in QoE research). In further research, it would de interesting to estimate the impact of P300 attenuation due to changes in quality and due to jittering effects. In the long run, it could be possible to utilize jittering effects between participants in order to measure an isolated neural response due to purely experimental quality variations.

In respect to frequency band analysis, the analysis of scalp potential distributions and a more detailed spectral analysis (lower and upper alpha band activation) will also be included. This will lead to an advanced differentiability between habituation effects (fatigue), hight mental workload, and changes in stimulus processing (impaired information processing).

In future research, the approach of measuring EEG frequency band power should not only be tested with different types of degradations, but also with different auditory, visual, and audio-visual stimuli. In the event that the results remain stable, the next question should be addressed, namely: When non-conscious activity is measured in the way described here, do these non-consciously processed degradations also result in changed customer behavior, e.g., in terms of shorter duration or a reduced frequency of voluntary media usage? In future studies, it could also be examined whether minor degradations (i.e., below the perceptual threshold) still have an effect on the emotional state of the user [116]. As user satisfaction is known to be an important factor for service acceptability, and as satisfaction is linked to the emotional state, answers to this question are considered to be important for telecommunication service providers, especially when they offer services which involve long-term media exposure (e.g., telephone conferencing, video-on-demand).

Another interesting field of future research are telephone-based spoken dialogue systems. Whereas the perceived quality of these systems can be assessed by using opinion tests [117], the developed test set-up could reveal more subtle differences and uncover influences on the cognitive state of participants. An optimized system using the test set-up could possibly ensure the satisfaction of customers.

Furthermore, tests with more natural profile patterns, including more frequent changes of bit rate, i.e., in the case of audio streaming, are needed. The test set-up should be extended such that the impact of several constant profiles on the vigilance level can be assessed; the hypothesis is that the total number of bits is not as important as the consistency of bit rate levels. In addition to the extended stimulus set-up, a post-questionnaire checking for reduced content perception due to impaired information processing should be incorporated.

Studies up to the present have concentrated on signals after stimulus presentation. Additionally, it would be interesting to determine if data—obtained by physiological measurement techniques—can enable better quality prediction on a single-trial basis. This could be done by using the neuronal signal preceding the stimulus for an estimation of the impact on perceived quality.

A similar approach to an experiment from Schubert et al. [118] could be implemented in this context. In this study, the frequencies preceding the stimulus were used to investigate the perceptibility of two competing somatosensory stimuli.

More specifically, the findings reported thus far have always focused on analyzing processes that occur within the participant *during* the perceptual and the descriptive event of quality assessment. While these methods can deliver useful additional information as shown, they do not take into account what impact the factors influencing QoE—such as the *cognitive state* of the listener—have on the judgment. Neuroscience studies prove that it is possible to detect not only the emotion and neuronal responses triggered by stimulus presentation, but that these methods can also be used to assess the general cognitive state *prior to presentation*. In simple words, precisely how the current state of a participant, be it emotional or cognitive, influences the process of forming a quality judgment could be measured. These results could lead to a better understanding of how the current state of participants influences subjective judgment.

The approach developed here can also be applied to interactive scenarios. As EEG data has also been successfully applied to audiovisual content [114], it would also be possible to test physiological responses evoked by audiovisual communications [119]. Besides the overall quality judgment, time-variant quality of audiovisual conversations could be considered as well [113]. New technological developments such as cloud gaming [120] and new web browsing approaches [121] also describe an interesting sphere for applying a combined subjective/physiological approach.

In addition to the ERP components already studied earlier, it would be interesting to know whether early components such as the N100 and P200 (N1/P2 amplitude reduction due to audiovisual synchrony) can be used as an additional measurement instruments during audiovisual conversation tests. In accordance with [122], the interrelation of asynchrony and audiovisual quality may have to be reconsidered, as it is likely they are not independent from one another.

In addition to classical media transmission technologies, physiological measurements such as alpha frequency band power can also be used in software applications that adapt automatically to the current cognitive state of users. One really demanding and necessary field for such applications is cognitive training for, e.g., dementia patients [123].

One research field that has not been extensively studied before in the context of Quality of Experience is the area of *brain stem* measurement, which will be a rewarding topic for future research. Basically, this method is similar to the *ERP* procedure and measures voltage differences due to neuronal activity. In contrast to ERP, this activity is produced by the brain stem and is different to the recorded ERP signal (i) temporally, as the signal emitted by the brain stem is measurable milliseconds after stimulus onset [124], and (ii) in terms of strength, which is much less. Especially worth mentioning is the work of the Nina Kraus group, that was able to prove that musical experience, for example, already has an influence information processing at the brain stem level [125]. It would be interesting to see whether quality expectations also come into play in this early phase of perception.

Another neurophysiological measure, *NIRS*, has been showing promising preliminary results in the domain of *QoE*-related research. Here, differences between oxygenated and deoxygenated, which is an indicator of neuronal activity, have been established. In a first experiment using auditory stimuli, significant correlations between recorded NIRS features and scored subjective ratings could be shown [126].

References

1. J.-N. Antons, R. Schleicher, S. Arndt, S. Möller, A.K. Porbadnigk, G. Curio, Analyzing speech quality perception using electroencephalography. IEEE J. Sel. Top. Signal **6**(6), 721–731 (2012)
2. J.-N. Antons, R. Schleicher, S. Arndt, S. Möller, G. Curio, Too tired for calling? A physiological measure of fatigue caused by bandwidth limitations, in *Proceeding of International Workshop on Quality of Multimedia Experience (QoMEX)* (2012), pp. 63–67
3. J.-N. Antons, K. Laghari, S. Arndt, R. Schleicher, S. Möller, D.O. Shaughnessy, T. Falk, Cognitive, affective, and experience correlates of speech quality perception in complex listening conditions, in *Proceeding of IEEE International Conference on Acoustics, Speech, and Signal Processing (ICASSP)* (2013), pp. 3672–3676
4. J.-N. Antons, F. Köster, S. Arndt, S. Möller, R. Schleicher, Changes of vigilance caused by varying bit rate conditions, in *Proceeding of International Workshop on Quality of Multimedia Experience (QoMEX)* (2013), pp. 148–151
5. J.-N. Antons, S. Arndt, R. Schleicher, S. Möller, Brain activity correlates of quality of experience, in *Quality of Experience*, ed. by S. Möller, A. Raake. T-Labs Series in Telecommunication Services, ch. 8, 1st edn. (Springer, Cham, 2014), pp. 109–119
6. European Commission—Eurostat: Statistical office of the European Union, *Mobile Phone Subscriptions—Per 100 Inhabitants (Database: tin00060)*. (European Union, Luxembourg, 2013)
7. ITU-T Recommendation G.722.2, *Wideband Coding of Speech at Around 16 kbit/s Using Adaptive Multi-Rate Wideband (AMR-WB)*. (International Telecommunication Union, Geneva, 2002)
8. ITU-T Recommendation H.264, *Advanced Video Coding for Generic Audiovisual Services*. (International Telecommunication Union, Geneva, 2013)
9. S. Möller, A. Raake, Motivation and Introduction, in *Quality of Experience* ed. by S. Möller, A. Raake. T-Labs Series in Telecommunication Services, ch. 1, 1st edn. (Springer, Cham, 2014), pp. 3 9
10. EN ISO 9241, *Ergonomic Requirements for Office Work with Visual Display Terminals (VDTs). Part 11: Guidance on Usability*. (International Organization for Standardization, Geneva, 1999)
11. K. Chen, C. Huang, P. Huang, C. Lei, Quantifying skype user satisfaction. SIGCOMM Comput. Commun. Rev. **36**(4), 399–410 (2006)
12. U. Jekosch, *Voice and Speech Quality Perception: Assessment and Evaluation* (Springer, Berlin, 2005)
13. ITU-T Recommendation P.10 (Amendment 3), *New Definitions for Inclusion in Recommendation ITU-T P.10/G.100*. (International Telecommunication Union, Geneva, 2011)

J.-N. Antons, *Neural Correlates of Quality Perception for Complex Speech Signals*, T-Labs Series in Telecommunication Services, DOI 10.1007/978-3-319-15521-0

14. S. Möller, *Assessment and Prediction of Speech Quality in Telecommunications* (Kluwer Academic Publishers, Dordrecht, 2000)
15. A. Raake, S. Egger, Quality and quality of experience, in *Quality of Experience*, ed. by S. Möller, A. Raake. T-Labs Series in Telecommunication Services, ch. 2, 1st edn. (Springer, Cham, 2014), pp. 11–33
16. M. Varela, L. Skorin-Kapov, T. Ebrahimi, Quality of service versus quality of experience, in *Quality of Experience*, ed. by S. Möller, A. Raake. T-Labs Series in Telecommunication Services, ch. 6, 1st edn. (Springer, Cham, 2014), pp. 85–96
17. ITU-R Recommendation BT.500-11, *Methodology for the Subjective Assessment of the Quality of Television Pictures*. (International Telecommunication Union, Geneva, 2002)
18. ITU-R Recommendation P.910, *Subjective Video Quality Assessment Methods for Multimedia Applications*. (International Telecommunication Union, Geneva, 2008)
19. ITU-T Recommendation P.800, *Methods for Subjective Determination of Transmission Quality*. (International Telecommunication Union, Geneva, 1996)
20. J.-N. Antons, S. Arndt, R. Schleicher, Effect of questionnaire order on ratings of perceived quality and experienced affect, in *Proceeding of International Workshop on Perceptual Quality of Systems (PQS)* (2013), pp. 1–3
21. A.R. Damasio, *Descartes' Error: Emotion, Reason, and the Human Brain* (G.P. Putnam, New York, 1994)
22. E.B. Goldstein, *Sensation and Perception* (Wiley, Hoboken, 2004)
23. H. Berger, Über das Elektroencephalogramm des Menschen. Eur. Arch. Psychiatry Clin. Neurosci. **87**(1), 527–570 (1929)
24. I. Miettinen, H. Tiitinen, P. Alku, P. May, Sensitivity of the human auditory cortex to acoustic degradation of speech and non-speech sound. BMC Neurosci. **11**(24), 1471–2202 (2010)
25. B. Lewcio, *Management of Speech and Video Telephony Quality in Heterogeneous Wireless Networks* (Springer, Heidelberg, 2014)
26. S. Möller, *Quality Engineering Qualität Kommunikationstechnischer Systeme* (Springer, Heidelberg, 2010)
27. J. Blauert, Analysis and synthesis of auditory scenes, in *Communication Acoustics*, 1st edn., ed. by J. Blauert (Springer, Heidelberg, 2005), pp. 1–25 (ch. 1)
28. U. Reiter, K. Brunnström, K. De Moor, M.-C. Larabi, M. Pereira, A. Pinheiro, J. You, A. Zgank, Factors influencing quality of experience, in *Quality of Experience*, ed. by S. Möller, A. Raake. T-Labs Series in Telecommunication Services, ch. 4, 1st edn. (Springer, Cham, 2014), pp. 55–72
29. S. Möller, M. Wältermann, M.-N. Garcia, Features of quality of experience, in *Quality of Experience*, ed. by S. Möller, A. Raake. T-Labs Series in Telecommunication Services, ch. 5, 1 edn. (Springer, Cham, 2014), pp. 73–84
30. M. Wätermann, *Dimension-based Quality Modeling of Transmitted Speech* (Springer, Heidelberg, 2013)
31. B. Weiss, D. Guse, S. Möller, A. Raake, A. Borowiak, U. Reiter, Temporal development of quality of experience, in *Quality of Experience*, ed. by S. Möller, A. Raake. T-Labs Series in Telecommunication Services, ch. 25, 1st edn. (Springer, Cham, 2014), pp. 367–381
32. J. Blauert, *Spatial Hearing: The Psychophysics of Human Sound Localization* (MIT Press, Cambridge, 1996)
33. ITU-R Recommendation BS.1534-2, *Method for the Subjective Assessment of Intermediate Quality Levels of Coding Systems* (International Telecommunication Union, Geneva, 2014)
34. ITU-T Recommendation G.107, *The E-model: A Computational Model for Use in Transmission Planning*. (International Telecommunication Union, Geneva, 2011)
35. ITU-T Recommendation P.863, *Perceptual Objective Listening Quality Assessment* (International Telecommunication Union, Geneva, 2011)
36. R. Schleicher, J.-N. Antons, Evoking emotions and evaluating emotional impact, in *Quality of Experience*, ed. by S. Möller, A. Raake. T-Labs Series in Telecommunication Services, ch. 9, 1st edn. (Springer, Cham, 2014), pp. 121–132

37. R. Parasuraman, The attentive Brain: issues and prospects, in *The Attentive Brain*, 1st edn., ed. by R. Parasuraman (MIT Press, Cambridge, 2000), pp. 3–17 (ch. 1)
38. S. Bech, N. Zacharov, *Perceptual Audio Evaluation: Theory Method and Application* (Wiley, Hoboken, 2006)
39. S.R. Quackenbush, T.P. Barnwell, M.A. Clements, *Objective Measures of Speech Quality* (Prentice Hall Englewood Cliffs, New Jersey, 1988)
40. F. Wichmann, N. Hill, The psychometric function: I. fitting, Sampling, and goodness of fit. Percept. Psychophysics **63**, 1293–1313 (2001)
41. J.-N. Antons, A.K. Porbadnigk, R. Schleicher, B. Blankertz, S. Möller, G. Curio, Subjective listening tests and neural correlates of speech degradation in case of signal-correlated noise, in *Proceedings of Audio Engineering Society Convention (AES)* (2010), pp. 1–4
42. A.K. Porbadnigk, J.-N. Antons, B. Blankertz, M.S. Treder, R. Schleicher, S. Möller, G. Curio, Using ERPs for assessing the (sub)conscious perception of noise, in *Proceeding of International Conference of the IEEE Engineering in Medicine and Biology Society (EMBS)* (2010), pp. 2690–2693
43. A.K. Porbadnigk, J.-N. Antons, M.S. Treder, B. Blankertz, R. Schleicher, S. Möller, G. Curio, ERP assessment of word processing under broadcast bit rate limitations. Neurosci. Lett. **500**(Supplement 1), e26–e27 (2011)
44. S. Arndt, J.-N. Antons, R. Schleicher, S. Möller, S. Scholler, and G. Curio, A physiological approach to determine video quality, in *Proceeding of IEEE International Symposium on Multimedia (ISM)* (2011), pp. 518–523
45. S. Scholler, S. Bosse, M.S. Treder, B. Blankertz, G. Curio, K.R. Müller, T. Wiegand, Towards a direct measure of video quality perception using EEG. IEEE Trans. Image Process **21**(5), 2619–2629 (2012)
46. C. Duncan, R. Barry, J. Connolly, C. Fischer, P. Michie, R. Näätänen, J. Polich, I. Reinvang, C. Petten, Event-related potentials in clinical research: Guidelines for eliciting, recording, and quantifying mismatch negativity, P300, and N400. Clin. Neurophysiol. **120**, 1883–1903 (2009)
47. D.A. Pizzagalli, Electroencephalography and high-density electrophysiological source localization, in *Handbook of Psychophysiology*, ch. 3, 3rd edn. ed. by J. Cacippio, L. Tassinary, G. Berntson (Cambridge University Press, Cambridge, 2007), pp. 56–84
48. M.S. Coles, M. Rugg, Event-related brain potentials: an introduction, in *Electrophysiology of Mind: Event-Related Brain Potentials and Cognition*, ch. 1, 1st ed. ed. by M.S. Coles, M. Rugg (Oxford University Press, Oxford, 1995), pp. 1–33
49. M. Fabiani, G. Gratton, K.D. Federmeier, Event-related potentials: methods, theory, and applications, in *Handbook of Psychophysiology*, ch. 4, 3rd edn. ed. by J.T. Cacioppo, L.G. Tassinary, G.G. Berntson (Cambridge University Press, Cambridge, 2007), pp. 85–119
50. G. Dornhege, J.R. del Millán, T. Hinterberger, D. McFarland, K.R. Müller, *Toward Brain-Computer Interfacing* (MIT Press, Cambridge, 2007)
51. B. Blankertz, R. Tomioka, S. Lemm, M. Kawanabe, K.R. Müller, Optimizing spatial filters for robust EEG single-trial analysis. IEEE Sig. Process Mag. **25**(1), 41–56 (2008)
52. K.R. Müller, S. Mika, G. Rätsch, K. Tsuda, B. Schölkopf, An introduction to kernel based learning algorithms. IEEE Neural Networks **12**(2), 181–201 (2001)
53. K.R. Müller, M. Tangermann, G. Dornhege, M. Krauledat, G. Curio, B. Blankertz, Machine learning for real-time single-trial EEG-analysis: from brain-computer interfacing to mental state monitoring. J. Neurosci. Methods **167**(1), 82–90 (2008)
54. B. Blankertz, M. Tangermann, C. Vidaurre, S. Fazil, C. Sannelli, S. Haufe, C. Maeder, L.E. Ramsey, I. Sturm, G. Curio, K.R. Müller, T. Wiegand, The Berlin brain-computer interface: non-medical uses of bci technology. Front. Neurosci. **4**, 198 (2010)
55. A.-N. Moldovan, I. Ghergulescu, S. Weibelzahl, C. Muntean, User-centered EEG-based multimedia quality assessment, in *Proceeding of International Symposium on Broadband Multimedia Systems Broadcasting (BMSB)* (2013), pp. 1–8
56. J. Perez, E. Deléchelle, On the measurement of image quality perception using frontal EEG analysis, in *Proceeding of International Conference on Smart Communications in Network Technologies (SaCoNeT)* (2013), pp. 1–5

57. S. Lal, A. Craig, Reproducibility of the spectral components of the electroencephalogram during driver fatigue. Int. J. Psychophysiol. **55**, 137–143 (2005)
58. Y. Punsawad, S. Aempedchr, Y. Wongsawat, M. Panichkun, Weighted-frequency index for EEG-based mental fatigue alarm system. Int. J. Appl. Biomed. Eng. **4**(1), 36–41 (2011)
59. J.A. Coan, J. Allen, Frontal EEG asymmetry as a moderator and mediator of emotion. Biol. Psychol. **67**, 7–50 (2004)
60. A. Gevins, M.E. Smith, Electroencephalography (EEG) in neuroergonimics, in *Neuroergonomics: The Brain at Work*, ch. 2, 1st edn. ed. by R. Parasuraman, M. Rizzo (Oxford University Press, Oxford, 2007), pp. 15–31
61. S. Lal, A. Craig, A critical review of the psychophysiology of driver fatigue. Biol. Psychol. **55**, 173–194 (2001)
62. S. Lal, A. Craig, Driver fatigue: electroencephalography and psychological assessment. Psychophysiology **39**, 313–321 (2002)
63. K. Hagemann, *The Alpha Band as an Electrophysiological Indicator for Internalized Attention and High Mental Workload in Real Traffic Driving*, PhD thesis (University Düsseldorf, Düsseldorf, 2008)
64. S. Arndt, J.-N. Antons, R. Gupta, K. Laghari, R. Schleicher, S. Möller, T.H. Falk, The effects of text-to-speech system quality on emotional states and frontal alpha band power, in *Proceeding of International IEEE/EMBS Conference on Neural Engineering (NER)* (2013), pp. 489–492
65. American Clinical Neurophysiology Society, Guideline 5: guidelines for standard electrode position nomenclature. J. Clin. Neurophysiol. **23**(2), 107–110 (2006)
66. R. Parasuraman, Neuroergonomics: bain-inspired cognitive engineering, in *The Oxford Handbook of Cognitive Engineering*, ch. 9, 1 edn. ed. by J.D. Lee, A. Kirlik, M. Dainoff (Oxford University Press, Oxford, 2013), pp. 159–177
67. E.A. Schmidt, M. Schrauf, M. Simon, A. Buchner, W.E. Kincses, The short-term effect of verbally assessing drivers state on vigilance indices during monotonous daytime driving. Transp. Res. Part F: Psychol. Behav. **14**(3), 251–260 (2011)
68. J.R. Jennings, P.J. Gianaros, Methodology, in *Handbook of Psychophysiology* ch. 34, 3rd edn. ed. by J. Cacippio, L. Tassinary, G. Berntson (Cambridge University Press, Cambridge, 2007), pp. 812–833
69. S. Luck, Ten simple rules for designing ERP experiments, in *Event-Related Potentials: A Methods Handbook*, ed. by T.C. Handy (MIT Press, Cambridge, 2005), pp. 17–32
70. S.H. Patel, P.N. Azzam, Characterization of N200 and P300: selected studies of the event-related potential. Int. J. Med. Sci. **2**(4), 147–154 (2005)
71. K.M. Spencer, Interpreting event-related brain potentials, in *Event-Related Potentials: A Methods Handbook*, ch. 1, 1st edn. ed. by T.C. Handy (MIT Press, Cambridge, 2005), pp. 3–16
72. M. Mustafa, S. Guthe, M. Magnor, Single trial EEG classification of artifacts in videos. ACM Trans. Appl. Percept. **9**(3), 1201–1215 (2012)
73. B. Blankertz, S. Lemm, M.S. Treder, S. Haufe, K.-R. Müller, Single-trial analysis and classification of ERP components—a tutorial. Neuroimage **56**, 814–825 (2011)
74. R. Näätänen, Mismatch negativity (MMN) as an index of central auditory system plasticity. Int. J. Audiol. **47**, 16–20 (2008)
75. M. Garrido, J. Kilner, K. Stephan, K. Friston, The mismatch negativity: a review of underlying mechanisms. Clin. Neurophysiol. **120**, 453–463 (2009)
76. L. Sculthorpe, D. Ouellet, K. Campbell, MMN elicitation during natural sleep to violations of an auditory pattern. Brain Res. **1390**, 52–62 (2009)
77. R. Näätänen, P. Paavilainen, T. Rinne, K. Alho, The mismatch negativity (MMN) in basic research of central auditory processing: a review. Clin. Neurophysiol. **118**, 2544–2590 (2007)
78. M. Pilling, Auditory event-related potentials (ERPs) in audiovisual speech perception. J. Speech Lang. Hear. Res. **52**, 1073–1081 (2009)
79. J.R. Folstein, C. Petten, Influence of cognitive control and mismatch on the N2 component of the ERP: a review. Psychophysiology **45**(1), 152–170 (2008)
80. J. Polich, Updating P300: an integrative theory of P3a and P3b. Clin. Neurophysiol. **118**(10), 2128–2148 (2007)

81. R. Näätänen, T. Kujala, I. Winkler, Auditory processing that leads to conscious perception: a unique window to central auditory processing opened by the mismatch negativity and related responses. Psychophysiology **48**, 4–22 (2011)
82. S. Koelsch, Music-syntactic processing and auditory memory: similarities and differences between ERAN and MMN. Psychophysiology **46**, 179–190 (2009)
83. K.M. Spencer, Averaging, detection, and classification of single-trial ERPs, in *Event-Related Potentials: A Methods Handbook*, ch. 10, 1st edn. ed. by T.C. Handy (MIT Press, Cambridge, 2005), pp. 209–227
84. M. Sokolova, N. Japkowicz, S. Szpakowicz, Beyond accuracy, F-score and ROC: a family of discriminant measures for performance evaluation, in *Proceeding of National Conference on Artificial Intelligence (AAAI)* (2006), pp. 1–6
85. S. Arndt, J.-N. Antons, R. Schleicher, S. Möller, G. Curio, Perception of low-quality videos analyzed by means of electroencephalography, in *Proceeding of International Workshop on Quality of Multimedia Experience (QoMEX)* (2012), pp. 284–289
86. L. Lindemann, S. Wenger, M. Magnor, Evaluation of video artifact perception using event-related potentials, in *Proceeding of ACM Applied Perception in Computer Graphics and Visualization (APGV)* (2011), pp. 53–58
87. H.-C. Li, J. Seo, K. Kham, S. Lee, Measurement of 3D visual fatigue using event-related potential (ERP): 3D oddball paradigm, in *Proceeding of 3DTV Conference* (2008), pp. 213–216
88. ITU-T Contribution COM 12–39, *Investigating the Subjective Judgment Process Using Physiological Data* (International Telecommunication Union, Geneva, 2013)
89. ITU-T Contribution COM 12–112, *Using Physiological Data for Assessing Variations of the Cognitive State Evoked by Quality Profiles* (International Telecommunication Union, Geneva, 2013)
90. ITU-T Recommendation P.810, *Modulated Noise Reference Unit (MNRU)* (International Telecommunication Union, Geneva, 1996)
91. R.C. Oldfield, The assessment and analysis of handedness: the Edinburgh inventory. Neuropsychologia **9**, 97–113 (1971)
92. A. Delorme, S. Makeig, EEGLAB: an open source toolbox for analysis of single-trial EEG dynamics. J. Neurosci. Methods **134**, 9–21 (2004)
93. A. Delorme, T. Mullen, C. Kothe, A.Z. Akalin, N. Bigdely-Shamlo, A. Vankov, S. Makeig, EEGLAB, SIFT, NFT, BCILAB, and ERICA: new tools for advanced EEG processing. Computat. Intell. Neurosci. **134**, 9–21 (2011)
94. R. Duda, P. Hart, D. Stork, *Pattern Classification* (Wiley, Hoboken, 2001)
95. J. Bortz, G.A. Lienert, K. Boehnke, *Methoden in der Biostatistik* (Springer, Heidelberg, 2008)
96. J. Bortz, *Statistik: Für Human- und Sozialwissenschaftler* (Springer, Heidelberg, 2005)
97. R. Schleicher, N. Galley, S. Briest, L. Galley, Blinks and saccades as indicators of fatigue in sleepiness warnings: looking tired? Ergonomics **51**(7), 982–1010 (2006)
98. A. Raake, *Speech Quality of VoIP: Assessment and Prediction* (Wiley, Hoboken, 2007)
99. D. Chan, A. Fourcin, D. Gibbon, B. Granstrom, M. Huckvale, G. Kokkinakis, K. Kvale, L. Lamel, B. Lindberg, A. Moreno, J. Mouropoulos, F. Senia, I. Trancoso, C. Veld, J. Zeiliger, EUROM—a spoken language resource for the EU, in *Proceeding of of the 4th European Conference on Speech Communication and Speech Technology* (1995), pp. 867–870
100. T. Halmrast, Sound coloration from (very) early reflections. J. Acoust. Soc. Am. **109**(5), 2303 (2001)
101. P. Rubak, Coloration in room impulse responses, in *Proceeding of Joint Baltic-Nordic Acoustics Meeting (BNAM)* (2004), pp. 1–14
102. T. Falk, C. Zheng, W.-Y. Chan, A non-intrusive quality and intelligibility measure of reverberant and dereverberated speech. IEEE Trans. audio Speech Lang. Process **18**(7), 1766–1774 (2010)
103. ITU-T Recommendation P.56, *Objective Measurement of Active Speech Level* (International Telecommunication Union, Geneva, 2011)

104. P.J. Lang, Behavioral treatment and bio-behavioral assessment: computer applications, in *Technology in Mental Health Care Delivery Systems*, ed. by J. Sidowski, J. Johnson, T. Williams (Ablex Publishing Corporation, New York, 1980), pp. 119–137
105. T. Falk, Y. Pomerantz, K. Laghari, S. Möller, T. Chau, Preliminary findings on image preference characterization based on neurophysiological signal analysis: towards objective qoe modelling, in *Proceedings of International Workshop on Quality of Multimedia Experience (QoMEX)* (2012), pp. 146–147
106. ITU-T Recommendation P.880, *Continuous Evaluation of Time Varying Speech Quality* (International Telecommunication Union, Geneva, 2004)
107. N. Johanneson, The ETSI computation model: a tool for transmission planning of telephone networks. IEEE Commun. Mag. **35**(1), 70–79 (1997)
108. M. Wältermann, I. Tucker, A. Raake, S. Möller, Extension of the e-model towards superwideband speech transmission, in *Proceeding of IEEE International Conference on Acoustics, Speech, and Signal Processing (ICASSP)* (2010), pp. 4654–4657
109. ITU-T Recommendation P.863, *Perceptual Objective Listening Quality Assessment* (International Telecommunication Union, Geneva, 2011)
110. ITU-T Recommendation P.862, *Perceptual Evaluation of Speech Quality (PESQ): An Objective Method for End-to-End Speech Quality Assessment of Narrow-Band Telephone Networks and Speech Codecs* (International Telecommunication Union, Geneva, 2001)
111. N. Cote, *Integral and Diagnostic Intrusive Prediction of Speech Quality* (Springer, Heidelberg, 2011)
112. ITU-T Recommendation P.862.2, *Wideband Extension to Recommendation P.862 for the Assessment of Wideband Telephone Networks and Speech Codecs* (International Telecommunication Union, Geneva, 2007)
113. B. Belmudez, B. Lewcio, S. Möller, Call quality prediction for audiovisual time-varying impairments using simulated conversational structures. Acta Acustica United Acustica **99**(5), 792–805 (2013)
114. S. Arndt, J.-N. Antons, R. Schleicher, S. Möller, G. Curio, Using electroencephalography to measure perceived video quality. J. Sel. Top. Sign. Process **8**(3), 366–376 (2014)
115. A.K. Porbadnigk, M.S. Treder, B. Blankertz, J.-N. Antons, R. Schleicher, S. Möller, G. Curio, K.-R. Müller, Single-trial analysis of the neural correlates of speech quality perception. J. Neural Eng. **10**(5), 1–20 (2013)
116. R. Schleicher, N. Galley, Continuous rating and psychophysiological monitoring of experienced affect while watching emotional film clips. Int. J. Psychophysiol. **46**(Suppl. 1), 51 (2009)
117. S. Möller, *Quality of Telephone-Based Spoken Dialogue Systems* (Springer, New York, 2005)
118. R. Schubert, S. Haufe, F. Blankenburg, A. Villringer, G. Curio, Now you'll feel it, now you won't: EEG rhythms predict the effectiveness of perceptual masking. J. Cogn. Neurosci. **21**(12), 2407–2419 (2009)
119. M. Vaalgamma, B. Belmudez, Audiovisual communication, in *Quality of Experience*, ed. by S. Möller, A. Raake. T-Labs Series in Telecommunication Services, ch. 14, 1st edn. (Springer, Cham, 2014), pp. 195–212
120. J. Beyer, S. Möller, Gaming, in *Quality of Experience*, ed. by S. Möller, A. Raake. T-Labs Series in Telecommunication Services, ch. 25, 1st edn. (Springer, Cham, 2014), pp. 367–381
121. D. Strohmeier, S. Egger, A. Raake, T. Hofeld, R. Schatz, Web browsing, in *Quality of Experience*, ed. by S. Möller, A. Raake. T-Labs Series in Telecommunication Services, ch. 22, 1st edn. (Springer, Cham, 2014), pp. 329–338
122. R. Eg, C. Griwodz, P. Halvorsen, D. Behne, Audiovisual robustness: exploring perceptual tolerance to asynchrony and quality distortion. Multimedia Tools Appl. **2014**, 1–21 (2014)
123. J. Gräske, H. Verbeek, P. Gellert, T. Fischer, A. Kuhlmey, K. Wolf-Ostermann, How to measure quality of life in shared-housing arrangements? A comparison of dementia-specific instruments. Qual. Life Res. **23**(2), 549–559 (2014)
124. R. Roeser, M. Valente, H. Hosfort-Dunn, *Audiology Diagnosis* (Thieme, Stuttgart, 2007)

125. P. Wong, E. Skoe, N. Russo, T. Dees, N. Kraus, Musical experience shapes human brainstem encoding of linguistic pitch patterns. Nature Neurosci. **10**, 420–422 (2007)
126. R. Gupta, K. Laghari, S. Arndt, R. Schleicher, S. Möller, D. O'Shaughnessy, T. Falk, Using fNIRS to characterize human perception of tts system quality, comprehension, and fluency: preliminary findings, in *Proceeding of International Workshop on Perceptual Quality of Systems (PQS)* (2013), pp. 1–6

Zeitfracht Medien GmbH
Ferdinand-Jühlke-Straße 7
99095 Erfurt, Deutschland
produktsicherheit@kolibri360.de